ABOUT THE AUTHOR

The creator and founding editor of *Entertainment Weekly* magazine and proprietor of Buzzmachine.com, one of the web's most popular and respected blogs about the internet and media, **JEFF JARVIS** also writes the new media column for *The Guardian* in London. In 2007 and 2008 he was named one of 100 worldwide media leaders by the World Economic Forum at Davos. He is on the faculty of the City University of New York Graduate School of Journalism in New York City.

What Would Google Do?

What Would Google Do?

**Reverse-Engineering the Fastest-Growing
Company in the History of the World**

Jeff Jarvis

HARPER
BUSINESS

NEW YORK ● LONDON ● TORONTO ● SYDNEY

HARPER
BUSINESS

A hardcover edition of this book was published in 2009 by Collins.

WHAT WOULD GOOGLE DO? Copyright © 2009 by Jeff Jarvis. All rights reserved. Printed in the United States of America. No part of this book may be used or reproduced in any manner whatsoever without written permission except in the case of brief quotations embodied in critical articles and reviews. For information, address HarperCollins Publishers, 10 East 53rd Street, New York, NY 10022.

HarperCollins books may be purchased for educational, business, or sales promotional use. For information, please write: Special Markets Department, HarperCollins Publishers, 10 East 53rd Street, New York, NY 10022.

FIRST HARPER BUSINESS PAPERBACK PUBLISHED 2011.

Designed by Level C

The Library of Congress has catalogued the hardcover edition as follows:

Jarvis, Jeff, 1954–
    What would Google do?/Jeff Jarvis.—1st ed.
        p. cm.
    Includes index.
    ISBN 978-0-06-170971-5
    1. Information technology—Management.   2. Technological innovations.   3. Creative ability in business.   4. Management.   5. Google.   I. Title.
    HD30.2.J375 2009
    658.4'012—dc22   2008040944

ISBN 978-0-06-170969-2 (pbk.)

11   12   13   14   15   WBC/RRD   10 9 8 7 6 5 4 3 2 1

For Tammy, Jake, and Julia

Contents

WWGD?   1

Google Rules   9

New Relationship   11
• Give the people control and we will use it
• Dell hell
• Your worst customer is your best friend
• Your best customer is your partner

New Architecture   24
• The link changes everything
• Do what you do best and link to the rest
• Join a network
• Be a platform
• Think distributed

New Publicness   40
• If you're not searchable, you won't be found
• Everybody needs Googlejuice
• Life is public, so is business
• Your customers are your ad agency

New Society   48
• Elegant organization

New Economy   54
• Small is the new big
• The post-scarcity economy
• Join the open-source, gift economy
• The mass market is dead—long live the mass of niches

• Google commodifies everything
• Welcome to the Google economy

New Business Reality   70
• Atoms are a drag
• Middlemen are doomed
• Free is a business model
• Decide what business you're in

New Attitude   82
• There is an inverse relationship between control and trust
• Trust the people
• Listen

New Ethic   91
• Make mistakes well
• Life is a beta
• Be honest
• Be transparent
• Collaborate
• Don't be evil

New Speed   103
• Answers are instantaneous
• Life is live
• Mobs form in a flash

New Imperatives   109
• Beware the cash cow in the coal mine
• Encourage, enable, and protect innovation
• Simplify, simplify
• Get out of the way

If Google Ruled the World   119

Media   123
• The Google Times: Newspapers, post-paper
• Googlewood: Entertainment, opened up
• GoogleCollins: Killing the book to save it

Advertising 145
• And now, a word from Google's sponsors

Retail 153
• Google Eats: A business built on openness
• Google Shops: A company built on people

Utilities 162
• Google Power & Light: What Google would do
• GT&T: What Google should do

Manufacturing 172
• The Googlemobile: From secrecy to sharing
• Google Cola: We're more than consumers

Service 182
• Google Air: A social marketplace of customers
• Google Real Estate: Information is power

Money 189
• Google Capital: Money makes networks
• The First Bank of Google: Markets minus middlemen

Public Welfare 199
• St. Google's Hospital: The benefits of publicness
• Google Mutual Insurance: The business of cooperation

Public Institutions 210
• Google U: Opening education
• The United States of Google: Geeks rule

Exceptions 222
• PR and lawyers: Hopeless
• God and Apple: Beyond Google?

Generation G 229

Afterword 243

Acknowledgments and disclosures 253

Index 257

What Would Google Do?

WWGD?

It seems as if no company, executive, or institution truly understands how to survive and prosper in the internet age.

Except Google.

So, faced with most any challenge today, it makes sense to ask: WWGD? What would Google do?

In management, commerce, news, media, manufacturing, marketing, service industries, investing, politics, government, and even education and religion, answering that question is a key to navigating a world that has changed radically and forever.

That world is upside-down, inside-out, counterintuitive, and confusing. Who could have imagined that a free classified service could have had a profound and permanent effect on the entire newspaper industry, that kids with cameras and internet connections could gather larger audiences than cable networks could, that loners with keyboards could bring down politicians and companies, and that dropouts could build companies worth billions? They didn't do it by breaking rules. They operate by new rules of a new age, among them:

- Customers are now in charge. They can be heard around the globe and have an impact on huge institutions in an instant.

- People can find each other anywhere and coalesce around you—or against you.

- The mass market is dead, replaced by the mass of niches.

- "Markets are conversations," decreed *The Cluetrain Manifesto*, the seminal work of the internet age, in 2000. That means the key skill in any organization today is no longer marketing but conversing.

- We have shifted from an economy based on scarcity to one based on abundance. The control of products or distribution will no longer guarantee a premium and a profit.

- Enabling customers to collaborate with you—in creating, distributing, marketing, and supporting products—is what creates a premium in today's market.

- The most successful enterprises today are networks—which extract as little value as possible so they can grow as big as possible—and the platforms on which those networks are built.

- Owning pipelines, people, products, or even intellectual property is no longer the key to success. Openness is.

Google's founders and executives understand the change brought by the internet. That is why they are so successful and powerful, running what The Times of London dubbed "the fastest growing company in the history of the world." The same is true of a few disruptive capitalists and quasi-capitalists such as Mark Zuckerberg, founder of Facebook; Craig Newmark, who calls himself founder and customer service representative—no joke—at craigslist; Jimmy Wales, cofounder of Wikipedia; Jeff Bezos, founder of Amazon; and Kevin Rose, creator of Digg. They see a different world than the rest of us and make different decisions as a result, decisions that make no sense under old rules of old industries that are now blown apart thanks to these new ways and new thinkers.

That is why the smart response to all this change is to ask what these disrupters—what Mark, Craig, Jimmy, Jeff, Kevin, and, of course, Google—would do. Google generously shares its own philosophy on its web site, setting out the "10 things Google has found to be true." They are smart but obvious PowerPoint lines helpful in employee indoctrination (especially necessary when your headcount explodes by 50 percent in a year—to 16,000 at the end of 2007 and to 20,000 before the end of the following year): "Focus on the user and all else will follow," Google decrees. "It's best to do one thing really, really well. . . . Fast is better than slow. . . . You can make money without doing evil. . . . There's always more information out there. . . . The need for information crosses all borders. . . ." These are useful, but they don't tell the entire story. There's more to learn from watching Google.

The question I ask in the title is about thinking in new ways, facing new challenges, solving problems with new solutions, seeing new opportunities, and understanding a different way to look at the structure of the economy and society. I try to see the world as Google sees it, analyzing and deconstructing its success from a distance so we can apply what we learn to our own companies, institutions, and careers. Together, we will reverse-engineer Google. You can bring this same discipline to other competitors, companies, and leaders whose success you find puzzling but admirable. In fact, you must.

Google is our model for thinking in new ways because it is so singularly successful. Hitwise, which measures internet traffic, reported that Google had 71 percent share of searches in the United States and 87 percent in the United Kingdom in 2008. With its acquisition of ad-serving company DoubleClick in 2008, Google controlled 69 percent of online ad serving, according to Attributor, and 24 percent of online ad revenue, according to IDC. In the U.K., Google's ad revenue grew past the largest single commercial TV entity, ITV, in 2008, and it is next expected to surpass the revenue of all British national newspapers combined. It is still exploding: Google's traffic in 2007 was up 22.4 percent in a year. Google no longer says how many servers its runs—estimates run into the millions—and it has stopped saying how many pages it monitors, but when it started in 1998, it indexed 26 million pages; by 2000, it tracked one billion; and in mid-2008 it said it followed one trillion web addresses. In 2007 and again in 2008, says the Millward Brown BrandZ Top 100, Google was the number one brand in the world.

By contrast, Yahoo and AOL, each a former king of the online hill, are already has-beens. They operate under the old rules. They control content and distribution and think they can own customers, relationships, and attention. They create destinations and have the hubris to think customers should come to them. They spend a huge proportion of their revenue on marketing to get those people there and work hard to keep them there. Yahoo! is the last old-media company.

Google is the first post-media company. Unlike Yahoo, Google is not a portal. It is a network and a platform. Google thinks in distributed ways. It goes to the people. There are bits of Google spread all over the web. About a third of Google's revenue—expected to total $20 billion in 2008—is earned not at Google.com but at sites all over the internet. Here's how they do it: The Google AdSense box on the home page of my blog, Buzzmachine.com, makes me part of Google's empire. Google sends me money for those ads. Google sends me readers via search. Google benefits by showing those readers more of its ads, which it can make more relevant, effective, and profitable because it knows what my site is about. I invited Google in because Google helps me do what I want to do.

I, in turn, help spread Google by putting its ads on my page and by embedding its YouTube videos, Google Maps, and Google search box on my blog. When I link to a page on the internet, I help Google understand what that page is about and how popular it is. I make Google smarter. With our clicks and links, we all do. Google is clever enough to organize that knowledge and take advantage of it. It exploits the wisdom of the crowd, and thereby respects us in the crowd. Google trusts us (well, most of us, except those damned spammers—but then Google has ways to ferret out the evil few among us). Google realizes that we are individuals who live in an almost infinite universe of small communities of interest, information, and geography. Google does not treat us as a mass. Google understands that the economy is made up of a mass of niches—that small is the new big. Google does not see itself as a product. It is a service, a platform, a means of enabling others that so far knows no limits.

As hard as it is to imagine today, Google could fail. It could grow too gangly to operate efficiently (I've heard rumblings from insiders that it's getting harder to accomplish things quickly because the company is just so huge). It could grow so dominant that government regulators try to break it up. In 2008, the U.S. Justice Department hired a top litigator to investigate Google's deal to serve ads on Yahoo and its dominance of the advertising market (though it should be noted that Google gained that position with the eager acquiescence of Yahoo, newspapers, and ad agencies). Google could also grow so big that it becomes hard to grow bigger; that's already becoming the case. Google could lose our trust the moment it misuses the data it has about us or decides to use our growing dependence on it as a chokehold to charge us (as cable companies, phone companies, and airlines do). It could lose its way or just screw up. When Gmail had a rare moment of dysfunction, Google CEO Eric Schmidt reminded the world, "We're not perfect."

So don't get hung up on trying to be Google, on mimicking what Google does. This book is about more than Google and its own rules and about more than technology and business. It's about seeing the world as Google sees it, finding your own new worldview, and seeing differently. In that sense, this isn't a book about Google. It's a book about you. It is about your world, how it is changing for you, and what you can gain from that.

It is hard to name an industry or institution—advertisers, airlines, retailers, auto makers, auto dealers, consumer-products brands, computer companies, fashion designers, telephone companies, cable operators, political candidates, government leaders, university educators—that should not be asking: What would Google do?

I will help you answer that question for your own world in the next section of this book, interpreting the wisdom of Google's ways as a set of rules to live and do business by in any sector of society. Then, in the following section, I'll illustrate how these laws can be applied across many companies, industries, and institutions, analyzing each as an exercise in thinking and acting differently. Finally, I examine how Googlethink is affecting our lives and the future of Generation Google. We begin by examining the new power structure in our economy and society, where we, the people, are suddenly in charge—empowered by Google.

Google Rules

New Relationship

Give the people control and we will use it
Dell hell
Your worst customer is your best friend
Your best customer is your partner

Give the people control and we will use it

Before getting to Google's laws, allow me to start with my own first law, learned on the internet:

Give the people control and we will use it. Don't, and you will lose us.

That is the essential rule of the new age. Previously, the powerful—companies, institutions, and governments—believed they were in control, and they were. But no more. Now the internet allows us to speak to the world, to organize ourselves, to find and spread information, to challenge old ways, to retake control.

Of course, we want to be in control. When don't you want to be the master of your work, business, home, time, and money? It's your life. Why would you cede control to someone else if you didn't have to? And once lost, wouldn't you take it back if given a chance? This empowerment is the reason we get so much angrier today when we are forced to wait on hold for computer service or at home for the cable guy or on the tarmac to get to our destination. It is why we lash out at companies—now that we can—on the web. But it is also why, when we are treated with respect and given control, we customers can be surprisingly generous and helpful.

Many good books have hailed the rise of the new, empowered customer. In this book, we ask: What should you do about it? How should this power-shift change the ways companies, institutions, and managers work? How do you survive? How do you benefit? The answer—the first and most important lesson in this book—is this: Companies must learn that they are better off when they cede control to their customers. Give us control, we will use it, and you will win.

Dell hell

Here is a case study in Jarvis' First Law involving Dell and me. But it isn't about me, the angry customer. It is about how Dell transformed itself from worst to first in the era of customer control. Dell had been the poster child for what you should not do. Then it became a model for what you should do.

After I quit my job as a media executive and left my expense account behind, I had to buy a new laptop. I bought a Dell, because it was inexpensive and because Dell had a reputation for good customer service. To be safe, I paid extra for at-home service.

From the moment I first turned on the computer, it had problems. I'll spare you the excruciating details of my shaggy laptop story. Suffice it to say that the computer had a number of bugs and I tried to fix them a number of times, spending countless hours on hold with people in faraway lands. Though I had paid for in-home service, I had to send the machine in to get it fixed, only to find something new wrong every time I got it back. Each time I dared to contact Dell, I had to start from square one: Sisyphus on hold. I never made progress. It drove me mad.

Finally, in hopeless frustration, I went to my blog in June 2005 and wrote a post under the headline, "Dell sucks." Now that's not quite as juvenile as it sounds, for if you search Google for any brand followed by the word "sucks," you will find the Consumer Reports of the people. I wanted to add to the wisdom of the crowd—which Google now made possible. I wanted to warn off the next potential customer who was smart enough to search for "Dell sucks" before hitting the buy button (which I should have done in the first place; the knowledge was there, at Google—all I had to do was ask). There were already a few million results for "Dell sucks." Mine was just one more. I didn't think I could fix my problem this way. I didn't think anything would come of it. But I got to vent steam. And that made me feel better. If I had known that my post would spark a popular movement and PR avalanche, I might have been more temperate in my language. But, hey, I was angry. This is what I blogged:

I just got a new Dell laptop and paid a fortune for the four-year, in-home service.

The machine is a lemon and the service is a lie.

I'm having all kinds of trouble with the hardware: overheats, network doesn't work, maxes out on CPU usage. It's a lemon.

But what really irks me is that they say if they sent someone to my home—which I paid for—he wouldn't have the parts, so I might as well just send the machine in and lose it for 7–10 days—plus the time going through this crap. So I have this new machine and paid for them to FUCKING FIX IT IN MY HOUSE and they don't and I lose it for two weeks.

DELL SUCKS. DELL LIES. Put that in your Google and smoke it, Dell.

Then something amazing happened. At first a few, then a score, then dozens and hundreds and eventually thousands of people rallied around and shouted, "What he says!" They left comments on my blog. They wrote blog posts elsewhere and linked to mine, spreading my story to thousands, perhaps millions more, and expanding Dell's anti–fan club. They emailed me, telling me their sad sagas in excruciating detail—and some continue to email me to this day.

The tale took on a life of its own as links led to more links and to a broader discussion about blogs, customers, and companies. We bloggers decided this was a test: Was Dell reading blogs? Was it listening? Houston Chronicle tech columnist Dwight Silverman did what reporters do: He called Dell to ask for its policy on blogs. "Look, don't touch," was the official reply. If customers want to talk to Dell, the spokeswoman said, they should talk to the company on its site, on its terms. But Dell's customers were already talking about Dell away from its site and control, on their own terms.

Soon, my blog posts were appearing progressively higher in Google search results for Dell, reaching the precious first page, only a few slots behind the link to Dell's home page. The conversation about my blog post was beginning to damage Dell's brand.

About this time, Dell's vital signs began falling. Customer-satisfaction ratings fell. Revenue results disappointed analysts. The share price dove, eventually losing half its value from about the time this saga began. That wasn't entirely my fault. I swear it wasn't. Though some have given me credit or blame for cutting Dell down to size, it's not true. I hardly did a thing. All I did was write a blog post that became a gathering point for many of my fellow frustrated Dell customers. They now stood beside me brandishing pitchforks and torches, brought together by the coalescing power of the internet, blogs, and Google. They were the people—not me—who should have been heeded by the company and by the analysts and reporters covering it. They told the real story of what was happening to Dell.

Two months after my Dell hell began, in August 2005, BusinessWeek told the tale in print. Under the headline, "Dell: In the bloghouse," the magazine wrote:

> PC industry circles have been buzzing in recent months that Dell's cus-
> tomer support is slipping—a claim bolstered on Aug. 16 by a University
> of Michigan study that showed a hefty decline in customer satisfaction
> from a year ago. So the last thing Dell needed was for someone to turn
> the customer-service issue into a cause célèbre.
>   Enter Jeff Jarvis.

About this time, I managed to get a refund for my laptop, though not as the result of blogging. I had sent an email to the company's head of marketing and, for snarky good measure, its chief ethics officer. The nice and patient lady whose job it is to talk to the irritants who get through to vice presidents called to offer help. She reached me on my mobile phone, I swear, just as I was in a computer store shopping for my Mac. She offered to exchange my computer for a new Dell laptop. I told her that I had lost trust in the company's products and services and just wanted my money back. She gave it to me.

And so, that August, I shipped the machine back and believed my Dell odyssey had ended. In what I thought was the final act in my silicon opera, I blogged an open letter to Michael Dell offering sincere and, I believed, helpful advice about bloggers and customers, who are more often now one and the same.

Your customer satisfaction is plummeting, your market share is shrinking, and your stock price is deflating.

Let me give you some indication of why, from one consumer's perspective . . . The bottom line is that a low-price coupon may have gotten me to buy a Dell, but your product was a lemon and your customer service was appalling. . . .

I'm typing this on an Apple PowerBook. I also have bought two more Apples for our home.

But you didn't just lose three PC sales and me as a customer.

Today, when you lose a customer, you don't lose just that customer, you risk losing that customer's friends. And thanks to the internet and blogs and consumer rate-and-review services, your customers have lots and lots of friends all around the world.

I told him about my fellow customers who'd chimed in with their complaints. I suggested he should have interns—better yet, vice presidents—reading what the world was saying about the company in the blogosphere. I also mentioned the big-time press, including BusinessWeek, that had picked up the story. Mocking Dell's own commercials, Fast Company magazine turned customer complaint online into a verb: "You got Dell'd."

But the tale I really loved, which I recounted in my open letter, came from Rick Segal, a blogging venture capitalist in Toronto who sat next to a couple of bank tellers in his office building's food court and heard them discussing the saga. That is how easily things spread online. Segal blogged the scene:

Lady one: "I was going to buy a new Dell but did you hear about Jeff Jarvis and the absolute hell he is going through with them?"
Lady two: "Yeah, I know, the IT guy told me that. . . ."

Segal had his own advice for Dell. "The pay-attention part: Lots of people (Dell?) are making the assumption that 'average people' or 'the masses' don't really see/read blogs so we take a little heat and move on. Big mistake." My advice for Dell continued with four simple tips:

1. Read blogs. Go to Technorati, Icerocket, Google, Bloglines, Pubsub, [search engines for blogs] and search for Dell and read what they're saying about you. Get it out of your head that these are "bloggers," just strange beasts blathering. These are

consumers, your marketplace, your customers—if you're lucky. They are just people. You surely spend a fortune on consumer research, on surveys and focus groups and think tanks to find out what people are thinking. On blogs, they will tell you for free. All you have to do is read them. All you have to do is listen.

2. Talk with your consumers. One of your executives said you have a look-don't-touch policy regarding blogs. How insulting that is: You ignore your consumers? You act as if we're not here? How would you like it if you gave someone thousands of dollars and they ignored you? You're not used to being treated that way. Neither are we. It's just rude. These bloggers care enough to talk about your products and service and brands. The least you can do is engage them and join the conversation. You will learn more than any think tank can ever tell you about what the market thinks of your products. But go to the next step: Ask your consumers what they think you should do. You'll end up with better products and you'll do a better job selling them to more satisfied customers who can even help each other, if you'll let them. It's good business, gentlemen.

3. Blog. If Microsoft and Sun and even GM, fercapitalismsake, can have their smartest [executives] blogging, so why shouldn't you? Or the better question: Why should you? Because it's a fad? No. Because it will make you cool with your kids? No. Blog because it shows that you are open and unafraid—no, eager—to engage your consumers, eye-to-eye.

4. Listen to all your bad press and bad blog PR and consumer dissatisfaction and falling stock price and to the failure of your low-price strategy and use that blog to admit that you have a problem. Then show us how you are going to improve quality and let us help. Make better computers and hire customer service people who serve customers.

"If you join the conversation your customers are having without you," I concluded, "it may not be too late." At last count, there were more than

600 responses to that blog post alone from fellow customers. One said: "I didn't know Dell had dropped the ball as far as quality was concerned. A few years ago, I would still be in the dark. The new grapevine is a great thing for consumers."

That was that, or so I thought. But eight months later, in April 2006, Dell began doing what I suggested and what others said would have been expensive and impractical: The company dispatched technical support staff to reach out to bloggers who had complaints, offering to solve problems, one at a time. Guess what happened: When technicians fixed bloggers' issues, Dell was rewarded with pleasantly surprised blog buzz. Bad PR turned good. Dell discovered that, contrary to what skeptics thought, this direct conversation with customers was an efficient way to learn about problems and solve them.

That July, Dell started its own blog, Direct2Dell. It got off to a rocky start, doling out promotion of the company and its products and not addressing the many elephants in its room. But after a few weeks, chief company blogger Lionel Menchaca entered the discussion with disarming directness and openness, linking and responding to Dell's critics and promising: "Real people are here and we're listening." He publicly discussed the case of an "infamous flaming notebook"—a computer whose battery exploded and caught fire rather spectacularly, pictures of which had sped around the internet (leading to a recall that also hit other computer manufacturers). He brought in other executives to be answerable to customers for ecommerce, product design, and, yes, customer service. The company dispatched staff to read blogs and comment on them. Later it enabled customers to rate and review products—positively and negatively—on Dell's site. Dell was listening and it was speaking in a new and credible human voice.

In February 2007, Michael Dell ordered the launch of IdeaStorm, a site where customers could tell Dell what to do, discussing and voting on the community's favorite ideas. There the company not only listened but acted. Customers wanted Dell to make computers for consumers with the open Linux operating system instead of Microsoft Windows. Dell's people fretted about problems that could arise if they installed one flavor of Linux versus another, but customers told them which way to go. Dell worried about supporting the new operating system, but customers said there was a community in place to handle that. Today, Dell sells Linux computers. In a later interview, Michael Dell acknowledged that selling Linux machines might not be a huge business, but it was an important symbolic act, the mark of a new partnership between company and customer.

I don't mean to take credit for Dell's transformation, only to note that Dell was now doing everything I had suggested in my open letter: reading and reaching out to bloggers, blogging itself, enabling customers to tell the company what to do, and doing it. So I had to give Dell credit: It was on the right road. Dell had joined the conversation.

The following April, I met Dell blogger Menchaca, who'd read on my blog that I was headed to Austin, in Dell's backyard, for a conference. He invited me out for beer with colleagues. On the way to the bar, Menchaca called his mother and told her that he was going to meet that blogger, Jeff Jarvis. Her response: "Are you sure you're going to be all right, dear?" My reputation had preceded me. But the Dell team came unarmed, as did I, and they convinced me that they had learned from the blogstorm around them and were using it to build a new relationship with their customers.

In the fall of 2007, I went to Dell headquarters in Round Rock, Texas, to interview Michael Dell for BusinessWeek and hear the company's turn-around story. As we sat down to talk, Dell wasn't exactly warm—that may just be the way he is (it's a CEO thing) or the problem could have been me (after all, I was the guy who'd raised hell). He began: "We screwed up, right?" He followed that confession with CEO bromides: "You gotta go back to the root cause and how to solve these things so they don't occur."

But eventually, Dell started to sound like a blogger himself. He might as well have had my first law etched in brass on his desk. "There are lots of lessons here for companies," he told me. "The simple way to think about it is, these conversations are going to occur whether you like it or not. OK? Well, do you want to be part of that, or not? My argument is, you absolutely do. You can learn from that. . . . And you can be a better company by listening and being involved in that conversation."

Of course, the company did more than blog to get itself out of trouble. Dell spent $150 million in 2007 beefing up its justifiably maligned customer-support call centers. Dick Hunter, former head of manufacturing, left retirement to head customer service and brought a factory-floor zeal for management and measurement to the task. The company had been judging phone-center employees on their "handle time" per call, but Hunter realized this metric only motivated them to transfer callers, getting rid of complaining customers and making them someone else's problem. Customers stood a 45 percent chance of being transferred; Hunter reduced that to 18 percent. More frightening, 7,000 of Dell's 400,000 customers calling each week suffered transfers seven times or more.

Instead of tracking "handle time," Hunter began to measure the minutes per resolution of a problem. Resolution in one call became the goal. He began a pilot program to reach out to 5,000 selected New Yorkers (if you can make it there . . .) before they had problems, hoping to replace brothers-in-law as their trusted advisers with a Dell expert. He insisted Dell could have direct relationships with at least half its 20 million customers.

At the same time, technicians were reaching out to bloggers to fix problems. More and more, I saw bloggers post amazed reactions when a published complaint led to contact from Dell and a solution. Adam Kalsey blogged about his problems reinstalling Microsoft's operating system in an old Dell machine and got immediate comment online from Brad, a Dell customer advocate, who fixed everything. Kalsey then blogged: "I'd heard from Jeff Jarvis that Dell was working hard to reverse their image of poor customer service. It's obvious that they're really trying to go the extra mile. . . . A year ago I recommended that a consulting client not buy Dell hardware (they did anyway). Now I couldn't imagine recommending anything else. Great work Dell and Brad." Group hug.

I asked the Dell team whether this approach was efficient, fixing problems one blog kvetch at a time. They insisted yes. When bloggers explained their problems, technicians could get right to the issue. Both the customer and the company saved time and money on the phone.

Dell's online PR turned around. After starting the program, by Dell's calculations, negative blog buzz dropped from 49 percent to 22 percent. That is, half the blog posts mentioning Dell had been negative before the outreach began; afterwards, only about a fifth of them were.

There are many lessons to be gleaned from Dell's saga: the danger of a mob forming around you in an instant if you treat your customers badly, the need to listen to and trust your customers, the benefits of collaborating with them, their generosity as a basis of a new relationship—all topics we will return to in subsequent chapters. But the primary lesson of Dell's story is this: Though we in business have said for years that the customer knows best and that the customer is boss, now we have to mean it. The customer is in control. If the customer isn't in control, there'll be hell to pay.

Your worst customer is your best friend

Now let's live out your worst nightmare—the day a blogstorm hits you—and see what you can learn from Dell to survive the crisis and emerge the better for it, having built a new relationship with your customers and the public.

Start at Google. Go there now, search for yourself—your company, your brands, even your own name—and find out what people are saying about you. If you haven't done it already, perform the same search at blog search engines Technorati, Icerocket, and Blogpulse, plus YouTube, Twitter (a blogging platform for short messages) and Facebook (where you may find groups formed for or against your company).

Now respond to people. Don't rely on an intern or a PR company to make the search and the contact. Do it yourself. Be yourself. Find someone who has a problem. Find out more about the problem by engaging in conversation. Solve it. Learn from it. Then tell people what you learned. You might have had such exchanges over the years via letters, phone calls, and underlings. But now the conversation will occur in public, as will your education. Don't be frightened. That's a good thing.

Let's say you find a customer—call him Angry Jim—who had a problem with your product—call it your eWidget. Jim writes on his blog that he got a lemon and shoddy service. He couldn't return it. The warranty was no help. He says in choice language that you don't give a damn about your customers.

Imagine all that Angry Jim could do online. He could complain on his blog and then start a site devoted to your problems—call it fWidget.com. As soon as he posts, a countdown starts as he and his readers wonder how long it will take you to notice and act. Jim may share the record of his interaction with your company, chronicling every phone call—including a log of hold time and what it cost him—and every automated, form-letter email. He can post audio of the calls, complete with repeated recorded reminders that his business matters to you. To spread his word, he will leave comments on related blogs and message boards and in Amazon reviews. He might make a YouTube video mashing up an eWidget commercial with his own message and jingle. If it's funny, it will spread. He can publish automated lists of other sites that are linking to him; this serves to gather his mob. Next, Jim could mobilize his fellow victims to take pictures of their busted widgets for Flickr. They could form a Facebook group devoted to complaining about eWidgets. When Jim finds an audience, his fWidget.com will rise on Google search results for eWidget. He's now competing to define your brand. It can't get worse but it does when a reporter calls asking about fWidget.com. Even if you don't listen to the conversation about you, reporters and competitors will. If you didn't think the problem was in the public before, you can be sure it will be now.

So what do you do? Run? Hide? Curse the lout? Sue him? Up your ad spending? Hire PR companies to just do *something* about this mess? Wait for it to go away? Look up your golden-parachute clause? You could try all that, but it won't do any good, not anymore. Your customers know where you are; you can't hide from them. Everything you and your employees do is being watched and made public in an instant. You have one chance to do the right thing, to rescue yourself. What will you do?

If I were you, I'd email Jim. Yes, he said nasty things about your widget. You may think he's an unreasonable complainer. You may fear that everything you say can and will be used against you in the court of public opinion (and you'd be right). You hate the idea of not being in control of this conversation. But remember: When you hand over control, you start winning.

Tell Jim that you want to understand the problem and fix it and that you're grateful for his help. He *is* helping you. He could just as easily have deserted you as a customer. Instead, he's telling you what went wrong and how to fix it. Keep in mind that if your employees had listened, things wouldn't have gotten this far. It escalated because Jim found himself talking to a brick wall with your brand on it. He wants to like your product; that's the reason he bought it. I'd draw as much knowledge, experience, and perspective out of customer Jim as I could—both because you will learn and because he will note that you are listening. Finally, I would encourage him to blog about the conversation and make it public (you won't have to invite him). Oh, and FedEx him a few new eWidgets for free.

Now comes the hard part. You have a company and a culture that are broken or this blogstorm would not have built up. Nobody gave a damn about your new buddy Jim, which really means that they didn't protect your reputation, brand, and business. I would call in all your C-people and project Jim's blog on the screen. Some execs will quibble with Jim: He voided his warranty; he called when it's the middle of the night in India; he didn't read the instructions; he's a complainer. But if Jim were a lone whiner, no mob would have gathered around him. His message rang true to too many customers.

Some executives will rely on reflexes: hiring consultants, making media appearances, updating the web site. Ignore them. It's time for new ways. Start by having your executives make the same searches you did, assigning their best people—nicest, most knowledgeable, most open—to solve every problem they find: repair, replace, or refund, whatever the customer wants. The cost is sure to be lower than the PR damage that could occur should the storm grow.

Next, I suggest you start a blog, where you openly and forthrightly share the problem and the solutions as they occur. I see no reason why a CEO should not open a direct conversation with the public. What's to fear from your own customers? Having set that example, the CEO can expect other executives and employees down the ranks to enter into the same conversation and learn from it. That will do more to change the culture—to finally make it customer-focused and mean it—than a dozen consultants, a hundred off-sites, or a million ad impressions.

Oh, and in that first blog post, don't forget to thank Jim.

Your best customer is your partner

Jim, no longer angry, will tell his blog friends about your turnaround. Having been heard, he will share more ideas about improving your products and company. Jim cares. He's not the enemy. He's a customer, even an advocate. Jim is your friend. Now the challenge—and opportunity—is to open the door to many Jims. The complementary challenge is to reorganize and reorient every division of the company—design, production, marketing, sales, customer support—around this new relationship with the people you used to call consumers but now should transform into partners.

Handing this new relationship over to one department—just customer service or PR or marketing—will not work. Outsourcing it to some crisis-management PR company or ad agency will make matters even worse. You have to transform your relationship with your public in every quarter of the organization. This new relationship—this partnership—should take over business-to-business companies, political campaigns, government agencies, universities, charities, any institution or enterprise.

To start, follow Dell's leads: blog, interact with bloggers, enable customers to critique your products, enable them to share ideas. Next, involve them in the genesis of your products, even your design process (an idea we will return to later in the chapter, "The Googlemobile"). In this hypothetical, why not take the next design of the eWidget—eWidget 2.0, of course—and make it public? Put it all out there: research, service reports, needs, design concepts, sketches, specifications, and new ideas. Go ahead, try it. The product is already in trouble. What could it hurt? I suppose your detractors and competitors might say that the eWidget is in such trouble, it means you're desperate. But that won't happen if your customers join the process with you, add value to the product, and take ownership of it. Then you'll get the last laugh.

You may extend this new relationship in many ways, such as inviting your customers to provide support, even marketing, and perhaps enabling customers to use your company as a platform to build their own companies. Through the rest of this book, we will return to the theme of this chapter—relationships—often. That is because the single greatest transformative power of the internet and Google has little to do with technology or media or even business. It's about people and making new connections among them. It all comes back to relationships.

The link changes everything

On the morning of September 11, 2001, I was on the last train into the World Trade Center from New Jersey, arriving just as the first of the terrorists' jets hit the north tower. Though I hadn't worked as a reporter for years, I was still a journalist and worked for a news company, so I decided to stay at what was clearly a big story—I didn't yet realize how big or how dangerous. I gathered notes on the scene and talked with survivors, calling my reports into my employer's news sites and newspapers. An hour later, I stood about a block from the edge of the World Trade Center site as the south tower collapsed. The cloud of destruction outran me. Blinded by the debris and covered in it, I was blessed to find refuge in a bank building. I then made my way on foot to Times Square, where I wrote my news story and finally, thank God, found my way home.

The next day, I had more to say about what I had seen and felt and the news around it, so I decided to start a blog. I had read blogs. I had also arranged my employer's investment in the company that started Blogger and popularized the form (it was bought by Google in 2003). I hadn't blogged myself, because I thought I had nothing to say. After 9/11, I did. So I planned to write the blog for a few weeks, until I ran out of memories.

But after writing my first posts, I learned a lesson that would change forever my view of media and my career; it would eventually lead to this book. A few bloggers in Los Angeles read what I had written, wrote about it on their blogs, and linked to me. I responded and linked to them. At that moment, a gong clanged over my head. I realized we were having a conversation—a distributed conversation, happening in different places at different times, which was made possible by the link. Soon enough, through Google's search, I could find other threads of the discussion around 9/11 and what I was writing. I saw a new structure of media: two-way and collaborative. I realized that this structure would redefine commerce, marketing, politics, government, education—the world. The link and search created the means to find anything and connect anyone. Now everyone could speak and all could hear. It enabled people to organize around any interest, task, need, market, or cause. The link and search started a revolution, and the revolution had only just begun.

Meg Hourihan, one of the creators of Blogger, wrote a groundbreaking essay in 2002, explaining the building blocks of this new system. (You can find it by searching Google for the title, "What We're Doing When We Blog.") Hourihan argued that the atomic unit of media online was no longer the publication or the page, with their old-media presumptions, but the blog post, which usually contains a discrete idea. Each post has a permalink, an address where it should be found forever so it can be linked to from anywhere. Hourihan realized that the permalink was both a means of organizing information and a way to build social networks on top of our distributed conversations. That is what happened when those bloggers in Los Angeles linked to my posts. We had a conversation, became friends, and even ended up doing business together. Our links connected us. "As with free speech itself," Hourihan wrote, "what we say isn't as important as the system that enables us to say it."

This system requires that everything about you, your product, your business, and your message has a place online with a permanent address so people can search and find you, then point to you, respond to you, and even distribute what you have to say. More than a home page, it's a home for every bit of what you do. Through what you put online, you will join with other people—friends, customers, constituents—in networks made possible by links, networks built on platforms such as Blogger and Google. You can now connect with people directly, without middlemen. The link and search are simple to use, but their impact is profound.

Do what you do best and link to the rest

The link changes every business and institution.

It's easiest to illustrate its impact on news. If the news business were invented today, post-link, everything about it—how news is gathered and shared and even how a story is structured—would be different. For example, in print, reporters are taught to include a background paragraph that sums up all that came before this article, just in case a reader missed something. But online, reporters can link to history rather than repeat it, because one reader might need to know more than a paragraph could impart whereas another reader, already informed, may not want to waste time on repetition. There are more uses of the link. When quoting from an interview, shouldn't a story link to the transcript or to the subject's site? If another news organization gets the only picture of a news event, shouldn't readers expect a complete story to link to it?

The link changes the structure and economics of a news organization. Every paper doesn't need its own golf writer when it's easier and cheaper to link to better tournament coverage at sports sites—freeing up resources that could be better used locally. Every paper doesn't need a local movie critic when movies are national and we are all critics. Papers should not devote resources to the commodified news we already know. They need to find new efficiencies, thanks to the link.

The link changes the structure of the industry. If a paper is going to stand out—if it wants people to find its content via search and links—then it needs to create stories with unique value. If they are to survive, newspapers must concentrate their resources where they matter, sending readers to others for the rest of the news. In short: Do what you do best and link to the rest.

Outside of media, retailers should link to manufacturers for product information. Manufacturers should link to customers who are talking about their products. Authors should link to experts (if only books enabled links). Headhunters, conferences, industry associations, and universities should use links to connect people who share needs, knowledge, and interests.

For almost every industry and institution, the link forces specialization. The notion of providing a one-size-fits-all product that does everything for everyone is a vestige of an era of isolation. Back then, Texans couldn't get the news directly from The New York Times, the Guardian, or the BBC, but today they can. Chicagoans couldn't buy great hot sauce in the local A&P, but now they can go online and buy it from HotSauce. com. These same pressures of specialization have killed generalist department stores—first with niche competitors in the mall and now with highly targeted retailers online. Serving masses, as we'll explore, is no longer the be all and end all of business. Serving targeted masses of niches—as Google does—is the future.

The specialization brought on by the link fosters collaboration—I'll do what I do and you'll fill in my blanks. It creates new opportunities to curate—when there are hundreds of lighting stores online or a thousand sites about Paris, there's a need for someone to organize them, linking to the best. And specialization creates a demand for quality—if you're going to concentrate on one market or service, you'd better be the best so people link to you, you rise in Google search results, and people can find and click on you.

In retail, media, education, government, and health—everything—the link drives specialization, quality, and collaboration, and it changes old roles and creates new ones. The link changes the fundamental architecture of societies and industries the way steel girders and rails changed how cities and nations were built and how they operated. Google makes links work. Google is the U.S. Steel of our age.

Join a network

Industries and institutions, in their most messianic moments, tend to view the internet in their own image: Retailers think of the internet as a store—a catalog and a checkout. Marketers see it as their means to deliver a brand message. Media companies see it as a medium, assuming that online is about content and distribution. Politicians think it is conduit for their campaign messages and fundraising (and a new way to deliver junk mail). Cable and phone companies hope the internet is just their next pipe to own.

They all want to control the internet because that is how they view their worlds. Listen to the rhetoric of corporate value: Companies *own* customers, *control* distribution, make *exclusive* deals, *lock out* competitors, keep trade *secrets*. The internet explodes all those points of control. It abhors centralization. It loves sea level and tears down barriers to entry. It despises secrecy and rewards openness. It favors collaboration over ownership. The once-powerful approach the internet with dread when they realize they cannot control it.

The internet adds networks of links over society, connecting people with information, action, and each other. It is in those connections that value is created, efficiency is found, knowledge is grown, and relationships are formed. Every link and every click is a connection, and with every connection, a network is born or grows stronger. That's how the internet spun its web, as the network of networks.

The more connections there are, the greater the value. You've certainly heard the old saw of network theory: One fax machine is worth nothing as it can talk to nothing, two are worth twice as much, and connecting millions of fax machines makes each one worth exponentially more (while the economies of scale—and the market for overpriced ink cartridges—also make each one cheaper to buy). The network is greater than the sum of its machines, but that's just a one-dimensional network: one machine talks to one machine one-way and once. The internet is a three-dimensional space of reciprocal links whose value multiplies with use and time. Google is the chief agent of that value creation.

Google performs this alchemy via search, of course: Seek and ye shall find anything you want in fractions of a second. Each time that happens—4.4 billion times a month in 2008 in the United States alone, according to Nielsen—another connection is made between a person and information or another person. Google creates a virtuous circle: The more we click on search results, the smarter Google gets; the smarter it gets, the better its results are, and the more we use Google.

Google supports its economy of clicks and links with ads, which appear on sites as small as my blog and as mighty as NYTimes.com; almost anyone can join its ad network. If Google thought like an old-media company—like, say, Time Inc. or Yahoo—it would have controlled content, built a wall around it, and tried to keep us inside. Instead, it opened up and put its ads anywhere, building an advertising network so vast and powerful that it is overtaking both the media and advertising industries even as it collaborates with and powers them online. There's Google's next virtuous circle: The more Google sends traffic to sites with its ads, the more money it makes; the more money those sites make, the more content they can create for Google to organize. Google also helps those sites by giving them content and functionality: maps, widgets, search pages, YouTube videos. Google feeds the network to make the network grow.

I am surprised that old media companies have not tried to copy Google's model—that is, creating open networks. But one new media company is building such a network. Glam is a web of women's sites covering fashion, health, celebrity, and more. In only two years, Glam grew to be the biggest women's brand online. As of this writing, it has more than 43 million users a month in the U.S. and more than 81 million worldwide according to comScore, surpassing the former queen of the hill, iVillage, with 18 million. iVillage, like Yahoo, operates under the old-media model: create or control content, market to bring in readers, and show them ads until they leave. Glam instead built a network of more than 600 independent sites, some created by lone bloggers, some by bigger media companies. Glam sells ads on those sites and shares revenue with them. Glam also replicates the best of the network's content at Glam.com, selling ads there—at a higher rate—and sharing that revenue, too. Glam gives its member sites technology and content to make them better. It gives them traffic and they give each other traffic, pointing to sister sites in the network. The more traffic each site gets, the more traffic it has to send around—that is the network effect, another virtuous circle. Glam also gives its sites prestige, for unlike Google, it is selective. Glam's editors find sites they like and highlight the best of the content, making Glam a curated content and ad network. That allows Glam to tell skittish advertisers that their messages will appear in a quality, safe environment, and advertisers will pay more for that.

There's another big advantage to Glam's network approach: cost. It need not hire expensive staff to create its wealth of content nor does it have to pay to license that content. At first, Glam guaranteed minimum payments to some sites—an investment that amounted to paying for content to get started—but it later eliminated those guarantees. Now it is a network of mutual benefit: The better content its sites create, the more traffic they get; the more traffic they can send around the network, the more Glam can sell ads at higher rates. Media companies should ask, WWGD? What would Glam do?

To be clear: Glam's no Google, at least not yet. It's not profitable and in 2008 was still taking venture capital—from, among others, the German publisher Burda—and investing in growth and technology. Its sites and their content can stand improvement. But I believe the model has legs and I'm no longer alone. The Guardian, Reuters, and Forbes each started blog ad networks to expand their content and advertising opportunities while their core businesses are challenged. These companies are taking a lesson from Google and its understanding of the networked architecture. I will argue later that restaurants and retail stores, certainly governments and universities, and even airlines and possibly insurance markets can operate like networks, creating more value when they create more connections in their worlds.

In 2005, I joined a roundtable held by the venture-capital firm Union Square Ventures in New York to talk about peer production and the creation of open networks and platforms. Counterintuitive lessons swirled around the room as entrepreneurs, investors, and academics analyzed the success of companies built this way. Across the table sat Tom Evslin, the unsung hero of the web who made the internet explode when, as head of AT&T Worldnet, he set pricing for unlimited internet access at a flat $19.95 per month, turning off the ticking clock on internet usage, lowering the cost for users, and addicting us all to the web.

Evslin gave a confounding lesson on networks. Explosive web companies—Skype, eBay, craigslist, Facebook, Amazon, YouTube, Twitter, Flickr, and Google itself—don't charge users as much as the market will bear. They charge as *little* as they can bear. That is how they maximize growth and value for everyone in the network. Evslin used an ad network to illustrate the value of building scale in this manner. An ad network that extracts the minimum commission it can afford out of ad sales for member sites will grow larger because more sites will join this network than its greedier competitors. Ad networks need a critical mass of audience before they can sell to top-tier advertisers, which pay higher rates. So charging less commission to grow larger can yield more ad sales at better prices.

It gets even more head-scratching: Evslin argued that if the company that runs the network is too profitable, it will attract competitors that will undercut it and steal market share. "If you're doing well but running at or close to breakeven," he explained later on his blog at TomEvslin.com, "you've made it impossible for anybody to undercut you without running at a deficit." To sum up Evslin's law of networks: Extract the minimum value from the network so it will grow to maximum size and value—enabling its members to charge more—while keeping costs and margins low to block competitors.

That's not how many old networks operate. Cable companies wrap their wires around us to squeeze maximum fees out. Ditto for phone companies, newspapers, and retailers. Charging what the market would bear made perfect sense for them. But now they face competition from next-generation networks. Skype—which at the end of 2007 had 276 million accounts in 28 languages—exploded as a free service before it added paid features that drastically undercut old phone companies. Its founders pulled value out of the business when eBay bought it. eBay itself had created a new retail marketplace by extracting little from each sale. Once eBay thought it was alone at the top, though, it started raising fees—but that allowed online retail competitors Amazon and Etsy to steal away merchants.

Evslin's poster child for network growth is craigslist. It foregoes revenue for most listings in most markets—charging just for job listings and for real estate ads in a few cities—and that made it *the* marketplace for most listings. "If Craig now attempted to maximize revenue by charging for a substantially higher percentage of ads, a door would be cracked open for competition," Evslin said. "There is no chance at current rates for a competitor to steal Craig's listings (and readers) by charging less." This is the economy in which Google operates. It had no revenue model for its first few years until it happened into advertising. "Bank users, not money," was Google vice president Marissa Mayer's advice on building new products and networks. She said in a 2006 talk at Stanford that Google doesn't worry about business models as it rolls out products. "We worry a lot about whether or not we have users." That is because on the web, "money follows consumers."

At the New York roundtable, an entrepreneur quoted legendary Israeli investor Yossi Vardi, who said that when he launched the pioneering instant-messaging service ICQ (later bought by AOL), he cared only about growing. "Revenue was a distraction," he decreed. This doctrine of growth over revenue was mangled in the web 1.0 bubble, when new companies spent too much of investors' money on marketing so they'd look big, only to collapse when money ran out and users vanished. Today's web 2.0 method for growth is to forgo paying for marketing and instead create something so great that users distribute it—it goes viral. Once it's big, then it can find the revenue. That money may not come directly from users in the form of fees or subscriptions but may come from advertising, ticket sales, merchandise sales, or from the value that is created from what the network learns—data than can be sold. I discuss such side doors for revenue later in the book.

Network economics may be confounding, but networks themselves are simple. They are just connections. You already operate in many networks. Go find the biggest whiteboard you can and draw your networks from various perspectives: First draw your company with all its relationships: customers, suppliers, marketers, regulators, competitors. Now draw a network from your customers' perspective and see where you fit in. Next draw your personal network inside and outside your company and industry. Draw your own company not as a boxy organizational chart but as a network with its many connections. In each, note where value is exchanged and captured (when you sell, you get revenue; when you talk with customers, you gain knowledge; when you meet counterparts, you make connections). Now examine how these networks can grow, how you can make more connections in each, how each connection can be more valuable for everyone. No longer see yourself as a box with one line up and a few lines down. Instead, put yourself in a cloud of connections that lights up each time a link is made, so the entire cloud keeps getting bigger, denser, and brighter—and more valuable. Then your world starts to look like Google's.

Be a platform

Networks are built atop platforms. The internet is a platform, as is Google, as are services such as photo site Flickr, blogging service WordPress.com, payment service PayPal, self-publishing company Lulu.com, and business software company Salesforce.com. A platform enables. It helps others build value. Any company can be a platform. Home Depot is a platform for contractors and Continental Airlines is a platform for book tours. Platforms help users create products, businesses, communities, and networks of their own. If it is open and collaborative, those users may in turn add value to the platforms—as IBM does when it shares the improvements it makes in the open-source Linux operating system.

Google has many platforms: Blogger for publishing content, Google Docs and Google Calendar for office collaboration, YouTube for videos, Picasa for photos, Google Analytics to track sites' traffic, Google Groups for communities, AdSense for revenue. Google Maps is so good that Google could have put it on the web at maps.google.com and told us to come there to use it, and we would have. But Google also opened its maps so sites can embed them. A hotel can post a Google Map with directions. Suburbanites can embed maps on their blogs to point shoppers to garage sales. Google uses maps to enhance its own search and to serve relevant local ads; it is fast becoming the new Yellow Pages. Google Maps is so useful on my iPhone that I'd pay for it.

In the old architecture and language of centralized, controlling businesses, Google Maps would be a *product* that *consumers* may use, generating an *audience* that Google could *sell* to advertisers. That's if Google wanted to stay in control. Instead, Google handed over control to anyone. It opened up maps so others could build atop them. This openness has spawned no end of new applications known as "mashups." In its June 2007 issue, Wired magazine credited Paul Rademacher, a DreamWorks animation programmer, with inventing the map mashup. In 2004, while looking for an apartment in the San Francisco area, he carried piles of printouts of craigslist ads and maps and thought—rather like the guy who first smeared peanut butter on chocolate—that they should be combined. He discovered he could dig into Google's code to put listings and maps together. After eight weeks, he had a demo that attracted thousands of users in a day. "I had no idea how big it would be. I just wanted to write something that was useful," he said. "Microsoft and Yahoo followed suit," Wired reported, "and before long the web was awash in map mashups." Google didn't sue Rademacher for messing with its product in an unauthorized manner. Google hired him.

Opening Google Maps as a platform spawned not just neat applications but entire businesses. Mobile phone companies are building Google Maps into their devices, which gets maps into the hands of new customers. Platial.com built an elegant user interface atop Google Maps that lets users place pins at any locations, showing the world anyone's favorite restaurants or a family's stops on vacation. Neighbors can collaborate and create a map pinpointing all the potholes in town. That map could, in turn, be embedded on a blog or a newspaper page. News sites have used maps to have readers pinpoint their photos during big stories, such as floods in the U.K.

Adrian Holovaty, a journalist/technologist—a rare breed the industry needs to clone—used Google Maps to make a news product and then a company. He took crime data from the City of Chicago and mashed it up with Google Maps, enabling residents to see every crime, by type, in any neighborhood. Because Holovaty's work was itself open, someone else mashed up his mashup, creating a site where commuters could trace their routes home and find all the crimes along the way. Holovaty folded his service, ChicagoCrime.org, into a new business, EveryBlock, which displays all sorts of data—from crime to building permits to graffiti cleanings—on neighborhood maps.

These new products and businesses were made possible because Google provided a platform. Businesses' use of the platform helped Google set the standard in mapping and local information. That gives Google huge traffic to its maps—tens of millions of users a month. Google invests to make maps better and better, licensing satellite pictures and hiring airplanes and cars to capture images of the ground. At the Burda DLD (Digital, Life, Design) conference in Munich in 2008, Google's Mayer, the vice president of search products and user experience, said, "We think of our geo technologies as building a mirror onto the world." She said Google Maps has coverage for half the world's population and a third of its landmass. The public's use of the maps adds yet more data, millions of bits of it. Users in Santiago, Chile, and Buenos Aires, Argentina, built the only comprehensive maps of their public-transit systems atop Google Maps. Users have also uploaded millions of geotagged photos associated with points on the maps, allowing us to get new views of places.

If you have a platform, you need developers and entrepreneurs to build on it, creating more functionality and value and bringing more users. Facebook did that. The social service got a big boost in attention and users when it enabled outsiders to create new applications inside the service. Within months, Facebook—which reached 500 employees in 2008—had 200,000 developers who created 20,000 new applications for users with virtually no staff cost to the company. When the service opened its Spanish and German versions, it didn't translate itself but created a platform for translation and handed the task over to users, who did the work for free.

Facebook profited because it expanded and users had more reasons to spend more time on the service. To do this, Facebook had to open up its infrastructure and some of its secrets to let outsiders program on its platform. By contrast, the European Union fined Microsoft $1.4 billion in 2008 because it failed to charge developers reasonable prices for access to its platform so they could build products on top of it.

Facebook went a step farther and killed some of the applications its internal programmers had written, believing the community would do a better job making them. My son and webmaster, Jake, who was 15 at the time, programmed his version of one of the apps Facebook killed, Courses, in which students share their class schedules. Pardon a moment's parental bragging, but his app rose to be No. 1 among its competitors—gathering information about 1.5 million classes—and he sold it to a competitor for enough to pay for a year of college.

Facebook did not charge Jake or other developers a penny for access to its code or its users, nor did Facebook take a cut of the advertising revenue developers earned. It was in Facebook's interest to help developers succeed because they helped the company grow in value. Grow it did, to the point that Microsoft made an investment in 2008 that valued Facebook at $15 billion (versus competitor MySpace's $580 million purchase by News Corp. in 2005).

I am a partner in a start-up called Daylife that created a platform to gather, analyze, organize, and distribute the world's news. Just as it was beginning, I took the founder, Upendra Shardanand, to meet venture capitalist Fred Wilson. At the end of our meeting, Wilson asked: "Can I use your platform to build my own business? And before you answer, let me tell you, the right answer is 'yes.'" Wilson sees building platforms as a strategic imperative. "In the economy we're in now, if you're not a platform, you'll be commoditized," he told me. Google will win against Microsoft and Yahoo, he argued, because too many companies have invested too much to build on Google's platform. That will make them loyal.

Questions to ask yourself: How can you act as a platform? What can others build on top of it? How can you add value? How little value can you extract? How big can the network atop your platform grow? How can the platform get better learning from users? How can you create open standards so even competitors will use and contribute to the network and you get a share of their value? It's time to make your own virtuous circle.

Think distributed

Most companies think centralized, and they have since the decline of the Sears catalog and the dawn of the mass market. Companies make us, the customers, come to them. They spend a fortune in marketing to attract us. We are expected to answer the siren call of advertising and trudge to their store, dealership, newsstand, or now, web site. They even think we want to come to them, that we are drawn to them, moths to the brand.

Not Google. Google thinks distributed. It comes to us whenever and however it can. Google's search box can appear on our browser or any page anywhere on the internet. If we do go to the trouble of traveling to Google.com's home page, we're rewarded with nothing but its spare search box and perhaps the occasional seasonal gag adorning the logo—but no ads. CNBC's Jim Cramer asked Google CEO Eric Schmidt in 2008 what the company could charge home-page sponsors. "Some number of billions of dollars," Schmidt said. But Google won't sell ads there because "people wouldn't like it." Yahoo and most other sites, on the other hand, try to make their home pages into destinations, crammed with content and advertising they believe will attract readers and serve marketers. Yet many users don't see these home pages. As many as 80 percent of a day's users at many news sites enter through search or links and never go to the home page.

Yahoo and many internet sites think of themselves as an end. Google sees itself as a means. Early in Yahoo's life, its cofounder, Jerry Yang, told me it was his job to get users in and out of Yahoo as quickly as possible. That changed when Yahoo decided to become a media company. Its new goal was to keep people inside its fence as long as possible. Years later, I heard Yang and his lieutenants brag about the "fire hose" of traffic they could bring from their home page. Like so many sites, they think the job of the home page is to take you where *they* want you to go. Google sees its home page as the way to get you to where *you* want to go. And when you get there, there's a good chance you'll find a Google ad or application. That is where Google wants to be: wherever you are.

Google distributes itself. It puts its ads on millions of web pages it does not own, earning billions of dollars for those sites and for itself. It offers scores of widgets—boxes of free, constantly updated content or functionality anyone can add to a web site or desktop: everything from weather to cartoons, chat to calendars, sports scores to photos, recipes to games, quotes to coupons. These widgets are filled with other companies' content; Google merely created the platform to distribute it. Yahoo, AOL, and other content sites should have created such distribution platforms years ago, cutting themselves up and offering their wealth of content and functionality to others to distribute and build upon. They didn't think that way. They didn't think distributed. They wanted to get us to come to them.

This understanding of the distributed web is what made Google buy YouTube for $1.65 billion in 2006 (though Google already had a lesser video service). YouTube grew to be the standard for video by making it easy not only to upload and play videos but also to embed—that is, to distribute—those videos on any site. My business partner, Peter Hauck, and I used this platform to build the blog Prezvid.com, which covered the 2008 U.S. presidential election through the eyes of YouTube. Google made it possible for us to create new content around its videos—promoting them—and we created a business by syndicating that content to WashingtonPost.com and CBSNews.com.

At a conference of media bigwigs in London in 2007, I got into an amicable debate about Google's model of distribution with my friend and former colleague Martin Nisenholtz, senior vice president for digital operations at The New York Times Company. I was urging the 200-plus worldwide media executives there to think like Google—that was the first time I publicly suggested that we should be asking, WWGD? I advised them to follow Google's example and distribute themselves as widely as possible, to go to where the readers are rather than make the readers come to them. Nisenholtz argued in response that some brands, such as The Times, are worth the trip to their site. He's right. But The Times is also worth distributing.

At that same conference, two brilliant consultants—Jeffrey Rayport of Harvard Business School and Andrew Heyward, former president of CBS News—advised the executives to turn their media properties "inside-out," inviting audiences in. They were half right. The problem with this formulation is that it still puts the media companies inside, at the center. That's not how their customers think of their worlds. People draw their me-spheres with themselves at the center and everyone else—especially those who want their money—on the outside. That's how companies and institutions should view themselves: on the outside, asking to come in.

"We can't expect consumers to come to us," is how the president of CBS Interactive, Quincy Smith, put it to The Wall Street Journal. "It's arrogant for any media company to assume that." Smith abandoned his network's strategy of creating a destination site where viewers would come to see its shows. He joked that the address for that failed portal should have been "CBS.com/nobodycomeshere." In its place, Smith developed a strategy built around the audience as the network, placing shows on as many sites and platforms as possible, making them embeddable, and hoping that people would distribute them farther. So far, it's working.

Enabling embedding gives networks something better than distribution. It gives them recommendations. If I put a clip of Jon Stewart's *Daily Show* on my blog, I'm recommending that you watch it. Even if I criticize the show, I'm saying there's something here worth seeing and discussing. You can watch it right then and there, without having to seek out Comedy Central.com. Such audience network strategies make viewers both distributors and marketers and get content into the wider conversation. When it works, viral distribution by the public can be more effective and certainly a lot cheaper than marketing to attract an audience.

The audience can also sell. In BarnesAndNoble.com's affiliate program, anyone can become a bookseller on a blog by adding a widget that recommends titles. If readers buy, the blogger earns a commission of 6 percent. It's no way to get rich but it does provide one more motivation for customers to become distributors and marketers. Bookstores are not alone. Search Google for any of many categories followed by the words "affiliate program" and you'll be surprised at the thousands that are happy to share revenue for selling gifts, flowers, shoes, insurance, Bibles, and, of course, porn. Retail, like search and content, can think distributed.

Newspaper classifieds were once the epitome of a centralized marketplace: You had to go to the paper to sell or buy a car or a home or find a job or an employee because that's where everyone did business. There was no other way for buyer and seller to find each other. Then came the internet and craigslist, whose founder, Craig Newmark, is blamed for sucking billions of dollars out of the newspaper industry. That's unfair. He simply created a tool that makes markets more efficient, leaving billions in the pockets of those doing the transactions. If Craig hadn't done it, someone else would have (no doubt Google wishes it had). craigslist is itself centralized. It's just a less expensive and more efficient marketplace. It's possible that more distributed solutions could supplant its database (though not its community). Specialized search engines such as SimplyHired.com, Indeed.com, and Oodle.com can aggregate every job posting and résumé around the web. The craigslist advantage, again, is that it does not charge what the market will bear—instead, as little as it will bear—and it has a loyal community.

I'm not arguing that everything online should stay distributed. When little bits of information and commerce spread out everywhere, they become hard to find. There is a need to aggregate them again—and a business opportunity there. Google News and Daylife (where I work) collect and organize headlines from all over the web so we can find all the latest news from anywhere. Some newspapers object to being aggregated. I believe papers should beg to be aggregated so more readers will discover their content. Daylife has taken headlines and put them in pages and widgets that sites can, in turn, distribute again. This pattern of distribution and aggregation is the yin-yang, push-pull of the distributed web: You want to be distributed, then aggregated, and then distributed again. You want to be found.

New Publicness

If you're not searchable, you won't be found
Everybody needs Googlejuice
Life is public, so is business
Your customers are your ad agency

If you're not searchable, you won't be found

Once upon a time, all roads led to Rome. Today, all roads lead from Google.

Google defines what your web presence should be. Of course, you need a web site. Who doesn't? But don't look at your site as a place where you get your message across. Don't obsess on a fancy home page and a path of navigation you want users to take (and please don't play music when I get there). Remember that many or most people won't see that home page. Most will likely come to you through Google after they ask a question.

The question is: Will you have the answer? That's how you should think of your site: answers for every question you can imagine, each on a page that is clearly and simply laid out so both Google and busy readers can find it and figure it out in an instant. If you're a manufacturer, customers should be able to find product details and support in an instant. If you're a politician, voters want to know your stands and record. If you're a food company, buyers want nutritional information. If you're a clothing company, shoppers want you to give the information a good sales clerk would—does this run large? Where can I buy your product? How do I contact you? Your users are already telling you what they want to know. Have your web folks show you the searches people made in Google when they clicked on a link to come to you. That is your starting list of questions to answer.

I learned about watching Google queries from About.com, the first media company made for the Google age. A vast majority of its traffic comes from Google. A large proportion of its ad revenue also comes from Google. About.com might as well be a division of Google, but it's not. It's merely built on Google's platform. About.com is owned by The New York Times Company, which bought it in 2005 for $410 million (and hired me to consult there). I'll confess I was dubious about the acquisition when it occurred, but I was wrong. Today, as papers struggle in the new economy, About.com is one of the rare bright spots in any newspaper company's P&L.

About.com at first wanted to compete with Google or even to be Google. Started by Scott Kurnit as The Mining Company in 1997—a year before Google was incorporated—its goal was to provide a human-powered guide to the internet. But as Yahoo also learned, that was hard and expensive, especially as the internet grew so unfathomably large. The company was rechristened About.com and became a content service with 700 sites maintained by independent writers and more than a million helpful, focused, and usually timeless articles about niche topics from car repair to thyroid disease. All these articles are structured so Google will find them easily.

About.com works hard to make itself Google-ready. Writers are taught search-engine optimization (SEO)—how to craft headlines, leads, page titles, and text around keywords so Google will recognize what each article is about. Writers are also taught to monitor search queries. If users are asking questions for which About.com doesn't have answers, they write articles with those answers. Keeping an eye on search terms is a preemptive readership survey, except instead of asking what people have read, About.com finds out what they want to read.

About.com's search-engine-optimization wizardry infiltrated its corporate sibling, The New York Times, where editors began to rewrite newspaper headlines for the web so Google's computers would understand them better and send more traffic to them. (For instance, the headline on a book review in the print Times may be clever but indecipherable unless you see the accompanying photo of the book cover and captions; online, the proper headline should include the title and author so anyone searching on either will find the review.) The Times also creates content aimed in part at pleasing Google: permanent topic pages on newsmakers and companies, which the paper hopes will become resources people will click on and link to over time, helping these pages rise in Google results, bringing in more traffic. Google was also a key reason why The Times changed its digital business model and stopped charging for content online (which I'll address in the chapter, "Free is a business model"). The most important benefit The Times received by opening up: Googlejuice.

Everybody needs Googlejuice

Googlejuice? That's the magic elixir you drink when Google values you more because the world values you more. It's another virtuous circle: The more links, clicks, and mentions you get, the higher you rise in Google's search results, offering you the potential for yet more clicks. The rich get richer, the Googley Googlier. I wonder whether, someday, companies will come to be valued not only on their revenue, marketshare, EBITDA, and profit but also on their Googlejuice.

The benefits of Googlejuice are lost on companies that do not make their information searchable—from local businesses that don't have sites to stores that don't post sales to manufacturers that don't publish product details to magazines that put content online in overcomplicated designs and databases that Google can't read. The benefits of search are also lost on a few media companies that resent Google and think they are punishing the big, bad beast by hiding from it. They're cutting off their noses to spite their faces. Various European papers have argued that Google and Google News are making money off their content and so they have demanded that Google stop searching their sites (which is easy for a site to do; just add a snippet of code to any web page to tell robots and spiders— the programs that crawl the web for search engines—to stay away). Blocking Google only means that it will stop sending readers, which is nothing short of suicide. That's like newspapers saying to a newsstand operator, "How dare you make a penny distributing my product? Give my papers back or I'll sue!" Google is their new newsstand.

It's insane to treat Google as the enemy. Even Yahoo doesn't (it asked Google to sell its ads). The goal today is to be Google's friend or at least, as adman Sir Martin Sorrell of WPP has dubbed Google, your "frenemy." The way to befriend and to exploit Google is to be searchable.

The way to become Google's enemy is to game and spam its search results. Evildoers will try to corrupt Google's algorithms to award their sleazy clients fraudulent Googlejuice. Some use automated software to create spam blogs—"splogs"—that carry fake content with lots of links to their clients, trying to trick Google into indexing and valuing all those links. Other companies use humans to do this dirty work, hoping to fake Google out and make it harder to ferret out the frauds. Some spammers pay people in poor nations pennies to create splog sites. And some companies hire bloggers to write nice things about their clients when, in reality, what they write is nothing any person would want to read. These often-unsuspecting bloggers are just creating more splog links to help give the bad guys more Googlejuice. It's insidious. Sadly, Google isn't always as diligent as it should be in cutting off the sploggers. Those pages also carry Google ads, which earn Google money.

What's good for big companies such as About.com is good for any small company or organization—or person. We all want to be found on Google. We all want Googlejuice. Customers now expect any information in the world to be available with one click. So every restaurant should have its menu, specials, hours, address, and more online. On a recent vacation, researching restaurants for the family, I went only to places that had web sites; I figured the others just didn't care enough. Not having an up-to-date web presence that Google can crawl and search and then present to users is like not having a phone number or a sign over the door. Today, that's particularly so because it's so easy to be on the web. The age of the geeky web priesthood is over. That restaurant can post its specials every day with a free weblog tool such as Blogger—which comes from Google. It can attract customers by buying ads on sites shown to people in the area—with Google. It can list itself on Google Maps and buy ads there, too.

The same can be said of you as an individual. You need a search presence. Your résumé should be online, because you never know when a job might come by. When you sell your house or car or golf clubs, you'll want them to be where they can be searched and found. As we'll discuss at the end of the book, without a Google shadow, old friends (and girlfriends and boyfriends) will never find you. Today, if you can't be found in Google, you might as well not exist.

How can you be sure to be found on Google? A new industry has emerged around just that need. Convention floors are filled with search-engine optimization companies promising to help you get to the promised land: the essential first page of search results for a topic relevant to what you do. Plenty of books and consultants can take you through all the technical details of searchability. I don't pretend to be a wizard of SEO, but there are a number of simple and obvious rules for how to think of your internet presence.

- Make sure every possible bit of information that anyone could want to know about you is on the web, searchable by Google.

- Construct information on pages so it can be understood by machine and man. In a word, be clear. If you're a dentist, say you're a dentist, not a smile doctor. Use the word "dentist" in the title of the page, the headline, and the beginning of what you write—make it so obvious even a computer couldn't be confused. This also means that when human beings come to the page, they'll know what you do. Clarity is always beneficial.

- Don't use fancy technology to make the content on your page dance and sing. Google won't recognize much of it (and readers will be irritated). Keep it simple.

- Don't bury your content inside fancy content management systems that stow it away in databases Google can't get to.

- Give everything you publish a permanent address—a permalink— so it can attract and accumulate more traffic and links and so Google has a place to which it can reliably send the people looking for you.

- Create separate pages for separate topics. If you're a restaurant, have a menu page and a directions page so, when I go searching for "Jeff's Chop House menu," Google can send me straight to your menu page.

- If there's any possible reason why anyone elsewhere on the web would want to link to you, make it easy for them to do so. If there are sites and bloggers writing about restaurants in your town, make them aware of your site. Google will notice their links, giving you a few more precious drops of Googlejuice.

• Once people come to your page, make sure you make it clear where they are: Put your brand on every page. When people go looking for an answer and find it via a click from a Google search, they often don't know where they have landed and who gave them their answer. Take credit.

Life is public, so is business

When the photo service Flickr started, its husband-and-wife founders, Caterina Fake and Stewart Butterfield, made a fateful if almost accidental decision. As Fake puts it, they "defaulted to public." That is, while other online photo services made the assumption that users would want to keep personal pictures private—stands to reason, no?—Flickr decided instead to make photos public unless told otherwise.

Amazing things happened. People commented on each other's photos. Communities formed around them. They tagged their photos so they could be found in searches because they wanted their pictures to be seen. They contributed more photos because they were seen. And as I will explain later, their usage of photos helped interesting ones to bubble up, which was possible only because they were all public.

Fake calls this condition "publicness," which is becoming a key attribute of society and life in the Google age. I believe publicness is also becoming a key attribute of successful business. We now live and do business in glass houses (and offices), and that's not necessarily bad.

Publicness is about more than having a web site. It's about taking actions in public so people can see what you do and react to it, make suggestions, and tell their friends. Living in public today is a matter of enlightened self-interest. You have to be public to be found. Every time you decide not to make something public, you create the risk of a customer not finding you or not trusting you because you're keeping secrets. Publicness is also an ethic. The more public you are, the easier you can be found, the more opportunities you have.

Your customers are your ad agency

For more than a century, the public face of companies has been their advertising, slogans, brands, and logos. How much better it would be if a company's public face were that of its public, its satisfied customers who are willing to share their satisfaction, and its employees who have direct relationships with customers. Brands are people.

If that's the ideal, then here's the goal: Eliminate advertising. Or at least fire your ad agency. Oh, you won't get rid of advertising entirely. You should be so lucky. But every time a customer recommends you and your product to a friend is a time when you don't have to market to that friend. It is possible today to think that one good word can spread as far as an ad would. This scenario is not hypothetical. When I had my problems with Dell, I could see them losing sales as people came to my blog and left comments saying they'd just decided not to buy a Dell, often adding that they'd told their friends their vow as well. There's no telling how much one pissed-off customer costs you today. The contrary is also true. A happy customer can sell your products. Now that bloggers are praising Dell online, new sales accrue as customers reconsider the company. When Dell started offering discounts to users of Twitter, who passed the word to more users, the company added $500,000 in sales in no time.

The more your customers take ownership of your brand, the less you will spend annoying people with your ads. I can hear your agency: You can't hand messaging over to the people; they'll be off-message. Well, tell your agency their message may be off. Your customers have always owned your brand.

Advertising is your last priority, your last resort, an unfortunate by-product of not having enough friends . . . yet. Learn this lesson from Google, which spends next to nothing on advertising. It became the fastest growing company in the history of the world without marketing. It grew thanks to its friends, not through ads. In its "10 things Google has found to be true," the company says its "growth has come not through TV ad campaigns but through word of mouth from one satisfied user to another." The generation that has that damned "Yahoo-ooo" sound stuck in their heads thanks to untold millions spent on commercials is the same generation that used and spread Google instead, for free.

Of course, Google's lucky. It created a spectacular product that solved a problem at just the right time, becoming essential to the internet and growing as it did with no limits on its scale. People need Google. They love Google. You may not be so lucky; you may be stuck selling a product that doesn't change the world in a market that's old and competitive. Sorry. But you may have great customer service and that's what people talk about. "Customer service is the new marketing," venture capitalist Brad Burnham blogged after having lunch with the best-known customer-service rep anywhere, Craig Newmark of craigslist. That law gained momentum as the title of a conference in 2008 held by GetSatisfaction.com, a company that created a platform for any customer to get help with any company. "Listening to our customers is actually the most perfect form of marketing you could have," said Mark Jarvis (no relation), chief marketing officer of Dell. Even if you don't have a product to love, you can still have a company worth admiring. Alloy Media surveyed college students in 2008 and found that 41 percent preferred socially responsible brands, a 24 percent increase in two years. Maybe that's why your customers will talk about you.

Once more, it comes down to relationships—relationships that are lived in public. Every time someone says something good about you online because of your product, service, reputation, honesty, openness, or helpfulness, you should knock another dollar off your advertising budget. Will it ever get to zero? Only if you're lucky.

Elegant organization

I sat, dumbfounded, in an audience of executives at the annual meeting of the World Economic Forum International Media Council in Davos, Switzerland, as the head of a powerful news organization begged young Mark Zuckerberg, founder of Facebook, for his secret. Please, the publisher beseeched him, how can my publication start a community like yours? We should own a community, shouldn't we? Tell us how.

Zuckerberg, 22 at the time, is a geek of few words. Some assume his laconicism is a sign of arrogance—that and his habit of wearing sandals at big business conferences. But it's not. He's shy. He's direct. He's a geek, and this is how geeks are. Better get used to it. When the geeks take over the world—and they will—a few blunt words and then a silent stare will become a societal norm. But Zuckerberg is brilliant and accomplished, and so his few words are worth waiting for.

After this publishing titan pleaded for advice about how to build his own community, Zuckerberg's reply was, in full: "You can't."

Full stop. Hard stare.

He later offered more advice. He told the assembled media moguls that they were asking the wrong question. You don't start communities, he said. Communities already exist. They're already doing what they want to do. The question you should ask is how you can help them do that better.

His prescription: Bring them "elegant organization."

Let that sip of rhetorical cabernet roll around on the palate for a minute. Elegant organization. When you think about it, that is precisely what Zuckerberg brought to Harvard—then other universities, then the rest of the world—with his social platform. Harvard's community had been doing what it wanted to do for more than three centuries before Zuckerberg came along. He just helped them do it better. Facebook enabled people to organize their social networks—the social graph, he calls it: who they are, what they do, who they know, and, not unimportantly, what they look like. It was an instant hit because it met a need. It organized social life at Harvard.

At this Davos meeting (which was off the record, but Zuckerberg gave me permission to blog it), he told the story of his Harvard art course. Zuckerberg didn't have time to attend a single class or to study. After all, he was busy founding a $15 billion company. The final exam was a week away and he was in a panic. It's one thing to drop out of Harvard to start a gigantic, world-changing company; it's another to flunk.

Zuckerberg did what comes naturally to a native of the web. He went to the internet and downloaded images of all the pieces of art he knew would be covered in the exam. He put them on a web page and added blank boxes under each. Then he emailed the address of this page to his classmates, telling them he'd just put up a study guide. Think Tom Sawyer's fence. The class dutifully came along and filled in the blanks with the essential knowledge about each piece of art, editing each other as they went, collaborating to get it just right. This being Harvard, they did a good job of it.

You can predict the punch line: Zuckerberg aced the exam. But here's the real kicker: The professor said the class as a whole got better grades than usual. They captured the wisdom of their crowd and helped each other. Zuckerberg had created the means for the class to collaborate. He brought them elegant organization.

Look at your constituents, customers, community, audience—even your competitors—and ask how you can bring them elegant organization, especially now, as the internet disrupts everything. Where some see a new world disorder, others see the opportunity to bring organization. This strategy is the foundation for so many internet companies: Google helps us organize around search, advertising, maps, documents, and more. Its mission, after all, is nothing less than to organize the world's information. eBay lets us organize markets for merchandise. Amazon helps us organize communities of consumer opinion around every product offered there. Facebook and other services like it—LinkedIn (big in business), Bebo (big in Europe), Google's Orkut (big in Brazil and India), and StudieVZ (big in Germany)—help us to organize our friends and colleagues. Skype, AOL, and Yahoo give us the tools to collaborate through chat, phone, and video, organizing our communication. Flickr lets us organize our photos and also communities of interest around them. del.icio.us does the same for our bookmarks and web recommendations. Daylife organizes the world's news. BlogAds lets bloggers organize ad networks. Wikipedia's platform enables us to organize our collective knowledge. Dell's support forums organize customers' knowledge. The internet brings us so many new paths to people, information, and functionality that we need help making sense of it. We've long needed help organizing ourselves. Government and media did that for us. Then internet portals and online media followed their centralized worldview. But the next generation of organizational enterprises—the Facebooks, Flickrs, and Wikipedias—don't organize us. They are platforms that help us to organize ourselves.

In his book *Here Comes Everybody*, New York University professor Clay Shirky argues that self-organization is a key to understanding the internet's impact on society. We can now organize without organizations. That is his law. Shirky studied the early years of Meetup, a New York company that uses internet tools to enable groups of people to get together in person. Its founder, Scott Heiferman, was inspired by Robert Putnam's book *Bowling Alone*, which argues that our communities are unraveling as we become more disconnected. Heiferman wanted to fix that by enabling groups to come together. "Use the internet to get off the internet," Meetup's home page urges. Where others saw disorder, Heiferman saw opportunity. In Shirky's examination of Meetup's first year, he learned that the groups that organized were not what you'd expect. The most popular? Not soccer moms or football fans or knitting circles but witches. Yes, witches. With reflection, this makes sense. Witches have so few ways to organize covens or coffee klatches. Meetup helped them do that.

When I ran newspaper sites, I tried to provide organization for communities with forum discussions and web-page tools, but I made the mistake of acting like a portal or media gateway: I decided what those communities were—parents, residents of a county, cooks. I thought I knew. If instead I had provided an open platform, who knows how many witches would have gathered in New Jersey? The key to offering elegant organization to individuals or groups—the key to all platforms—is to enable others to use the tool as they wish. They know their needs. Such openness and flexibility also enables more groups to form. Each one may be small, but altogether, they add up to a larger network of groups—a mass of niches.

There is an ongoing debate about who will win the social space, what company will own the social web. That's a wrong-headed view of the opportunity. The internet already is a social network. So is life. The internet merely provides more means to make more connections. The winner is not the company that gets us to come in and be social inside a wall: the social AOL or MySpace or, for that matter, Facebook. The winner will be the one that figures out how to bring elegant organization to the disorganized social network that the internet already is. We are waiting for the Google of people. Zuckerberg's stated ambition is to be that next Google. And Google is afraid that he might succeed, which is why it created a standard called Open Social and banded together with other social networks, hoping to beat Facebook at its own game. To win, Facebook needs to be more open, to look beyond its walls and figure out how to take its organization to the rest of our lives online. I'll bet they will be smart enough to do it.

Politics is at last learning the skills of self-organization. In 2004, Howard Dean's presidential campaign used blogs and discussion as well as in-person Meetups to organize volunteers and raise money. Barack Obama's 2008 campaign made brilliant use of social tools, including Facebook and the iPhone, to organize rallies and rake in donations. More profound, it used the social web to organize a movement. It also took advantage of the fact that other communities—such as that inside the DailyKos blog—had gathered around Obama. It didn't hurt that one of Facebook's founders, Chris Hughes, was an adviser to Obama's campaign.

We want to be connected. In the internet age, we have gained a reputation for being antisocial, for sitting on our couches, laptops on laps, earphones on ears, never talking to anyone. But in truth, we're talking to more people from more places more often than ever before because we have more ways to do it. Thanks to Google and Facebook, I've reconnected with old colleagues and friends and made new business connections. The success of Facebook comes in great measure from returning us to real identities, real reputations, and real relationships. Anonymity had its place on the internet—it was fun for awhile, when, as the legendary New Yorker cartoon says, nobody knew you were a dog. But now we're settling back to our norm: hanging out with people we know, like, and trust. We often want to do more than hang out together: We want to accomplish things together.

Organization is a business model. Look at the communities around you—not communities you start but communities you serve. There is one, even if you are an airline or a cable company or a doctor's office. There is a community of people with like interests and needs. Have you enabled them to talk, to share what they know and need to know, to support each other, to do business together, even to socialize? You are probably working with a group of people who have shared concerns: Staples customers who run small businesses, Gourmet readers who like to go on food holidays, Cisco router buyers who know a lot about networks, students who need jobs, alumni who are hiring. They are gathered outside your house. All you have to do is open the windows all around to let them talk with each other.

But do be careful. Don't assume these people care about you or think of themselves as members of your community. Don't think that you can create a community. They're not yours. They're not going to start wearing Target T-shirts or singing the Toyota song—not unless you have an extraordinary product and brand (such as an entertainment brand or a hot designer label or Apple). That's about the silliest thing I hear from any company: They talk about *their* community. I have sat in meetings with major consumer brands—candies, soaps, stores—as they say that they have communities that will come to their sites and do what they think they should do. Remember Zuckerberg's advice: Communities are already doing what they want to do. If you're lucky, they'll let you help them.

Once a community does gather around you, be aware that you don't own it; the community owns itself. American Girl, the doll brand, started an online club as a safe place where young girls could communicate with each other and play games to earn points and gifts. The business wasn't big enough for owner Mattel, so one day it up and killed the club, crushing my daughter, Julia, and cutting her off from the friends she had made there. Mattel should have learned who runs its town. It's a lesson Barack Obama learned when his followers, disappointed with his stand on an issue, used his own campaign platform to organize a protest against him. Once you hand over control, you can't take it back.

We no longer need companies, institutions, or government to organize us. We now have the tools to organize ourselves. We can find each other and coalesce around political causes or bad companies or talent or business or ideas. We can share and sort our knowledge and behavior. We can communicate and come together in an instant. We also have new ethics and attitudes that spring from this new organization and change society in ways we cannot yet see, with openness, generosity, collaboration, efficiency. We are using the internet's connective tissue to leap over borders—whether they surround countries or companies or demographics. We are reorganizing society. This is Google's—and Facebook's and craigslist's—new world order.

New Economy

Small is the new big
The post-scarcity economy
Join the open-source, gift economy
The mass market is dead—long live the mass of niches
Google commodifies everything
Welcome to the Google economy

Small is the new big

Mind you, big is still big. Wal-Mart is the largest company on earth. Big-box stores such as Home Depot continue to drive mom-and-pop hardware shops out of business. Media companies are conglomerating. Airlines are merging. Even small churches are being turned into condos thanks to the rise of megachurches. The Super Bowl can still draw 97 million viewers. Hell, Google itself isn't just big; it's ginormous. No, big won't go away.

But small is rising. A tiny start-up can become a manufacturing company using somebody else's factory and distribution while selling to a worldwide market that can find its products via Google. Any of us can start a highly specialized and targeted media company using blog software and paying for it with Google ads. One person can plant a seed to start a political movement.

There won't be a single new retail behemoth to battle Wal-Mart like Japanese monsters in Tokyo Bay. Instead, Wal-Mart and other big chains are getting nipped at their heels by a million tiny competitors—a half a million of them on eBay alone. In 2007 eBay sold $59.4 billion in merchandise from 547,000 online stores. It may be dwarfed by Wal-Mart's $345 billion, but in 2007 eBay beat the sales of America's largest department-store chain, Federated (aka Macy's), with revenues of $26.3 billion in 853 stores.

Some weblogs now have more traffic and links than major media sites. Gawker Media, a gaggle of gossipy blogs started by Nick Denton, boasted in July 2008 that its dozen sites had double the web traffic of the Los Angeles Times online—254 million vs. 127 million page views in a month. All weblogs, as a group, now have an audience of readers (57 million as early as 2006, according to the Pew Internet and American Life survey) that is larger than the number who buy daily newspapers (50 million in early 2008, according to the Newspaper Association of America). Even more striking, Pew said back in 2004 that 53 million Americans had used the internet to "publish their thoughts, respond to others, post pictures, share files and otherwise contribute to the explosion of content available online." The writers are starting to outnumber the readers.

The Lilliputians have triumphed. The economies of scale must now compete with the economies of small. What changed is the definition of "big enough"—big enough to make money, big enough to survive and succeed. The tipping point of critical mass in business has fallen from the sky to eye level. Once upon a time in retail, you had to have a store, which needed location, location, location; capital to fill it with inventory; and cash flow to hire staff and buy ads to bring in customers. Then you had to have a chain of stores to gather muscle with suppliers and create marketing efficiencies. Now, you can find customers via eBay, Amazon (which is as much a platform for retailers as it is a retailer itself), Google (where you can buy inexpensive and targeted ads), and new online marketplaces of neat and unique stuff such as Etsy.com (which sells handmade clothes and crafts). Profits accrue sooner because you don't own bricks or necessarily stock inventory or spend a fortune on marketing.

Once upon a time, you couldn't write for a living unless you were paid by a big publisher, the guy who could afford to own printing presses because he was the guy who had the big audience (a virtuous circle of its time). Now many writers make money blogging. Enough money? Well, that's up to you. It could be enough to pay for your internet hosting or maybe a lunch or two—or a decent living. Here's an accounting of the value of my blog: In 2007, I made $13,855 in ad revenue ($4,450 of that from Google) on Buzzmachine. I shouldn't have quit my day job, you say. But Buzzmachine is what got me appointed as a journalism professor at the City University of New York Graduate School of Journalism (worth not quite six figures a year) and consulting and speaking gigs (worth a few times that in good times) and the contract for this book (worth about double those gigs). So over a few years, my weblog is easily worth seven figures. My cost: $327 a year for deluxe internet hosting. There are bloggers who make—and whose blogs are worth—much more. But Buzzmachine is big enough.

Calculate in the falling cost of work if you want to go it alone online—no office, no commuting, no suits—and our definitions of profitability, critical mass, and success all shrink. The cost of independence has dropped. In an age when so many people are sick of their jobs—you know who you are—this self-reliance is empowering. Loyalty from employer to employees died in my lifetime. Now, given the chance to earn FU money and leave office politics behind, there is less loyalty from employees to employers as well. We'll see more people trying to make it on their own because they want to and they can—or because they have no choice when shrinking companies lay them off.

What should their former employers' relationships be to these newly independent agents? Companies should encourage and support some one-man spin-offs. After U.K. football writer Rick Waghorn was laid off from his paper in Norwich, he started his own football blog and community with a former business colleague. Their old paper viewed them as competition. Foolish. It was the paper that had built Waghorn's brand and audience. When it fired him, it lost that investment along with his content. It didn't have to. Instead, the paper should have sold Waghorn's ads and promoted his site. It could have taken advantage of his expertise, work, reputation, and audience without having to pay his salary. Meanwhile, Waghorn would have been able to build a company. Everybody won. If I were to run the paper, I'd invest in Waghorn. I'd build a network of Waghorns.

But it's not easy being a Waghorn. Without the paper acting as his promoter, he and others like him have a hard time building the critical mass of audience and advertisers they need. Even in the small-is-the-new-big era, it is possible to be too small. At a conference on collaborative journalism I ran at CUNY, online news entrepreneur Mark Potts said that perhaps the only way to succeed at being small is to be part of something big: a network. Big still has its place. It's the relationship between small and big that is evolving.

If we end up with more independent agents able to do what they do best—making jewelry or providing computer advice or writing—I hope we will start to see a reversal of the malling of the world that big manufacturing and retail have brought: the sameness of scale. Before the Berlin Wall fell, I was amazed to find a Benetton store even in communist East Berlin. They were everywhere. Starbucks cafés and Pret A Manger sandwich shops (which are one-third owned by McDonald's) have replaced pubs all around London. Hip Soho in New York is filled no longer with artists and boutiques making singular merchandise but with Banana Republics. Everything's the same; nothing's unique; and that takes the fun out of making, buying, and owning. The small-is-the-new-big world could bring variety back. The craftsman lives again on Etsy, eBay, Amazon, and hip T-shirt company Threadless (where the buyers and wearers make the designs).

In 2005, I read two posts by marketing visionary, author, and blogger Seth Godin about companies that just didn't care. He inspired me to blog that we could now create new competitors. "Small is the new big," I wrote. At the same moment, Godin, similarly inspired, wrote the same line on his blog (and he beat me to using it as the title of a book). "Get small," Godin blogged. "Think big."

The post-scarcity economy

We are entering a post-scarcity economy in which Google is teaching us to manage abundance, challenging the bedrock rule of economics, first written in 1767: the law of supply and demand.

Many industries built their value on scarcity. Airlines, Broadway theaters, and universities had only so many seats, which meant they could charge what they wanted for them. They were scarce and thus more valuable. Newspapers owned the only printing press in town and you didn't, so they could charge you a fortune to reach their audience. Shelf space in grocery stores was limited, so manufacturers paid for the privilege of selling their boxes there. Television networks had a finite number of minutes in the day with only so many eyeballs watching, so advertisers competed to buy their commercial time. Scarcity was about control: Those who controlled a scarce resource could set the price for it.

Not anymore. Want to sell your product to a targeted market? You don't need to fight for a spot on the shelf in 1,000 stores; you can now sell to anyone in the world online. Looking for a dress everyone else doesn't have when everyone else shops in the same mall? Today you can find no end of choice only a click and a UPS delivery away. Don't want to buy The New York Times on the newsstand or pay for access to WSJ.com for news on your industry? Now there are countless sources of the same information. Even if The Journal reports a scoop behind its pay wall, once that knowledge is out—quoted, linked, blogged, aggregated, remixed, and emailed all over—it's no longer exclusive and rare. It's no longer possible to maintain that scarcity of information.

Advertising agencies act as if ad inventory were still scarce, though online there is a virtually unlimited supply of advertising opportunities now. Agencies have always liked one-stop shopping. Every fall, they go to network upfront parties, where shows are previewed, wine is poured, and much of the entire season's ad inventory is sold off. Prime slots such as Thursday nights—when studios advertise weekend movie premieres—sell out at ever-higher prices even though the audience watching broadcast TV is getting ever-smaller (and, goes the reasoning, scarcer). Just as nobody gets fired in technology for buying IBM, according to the old business rule, nobody gets fired in advertising for buying TV. Agencies' willful ignorance of new ad economics is a product of their own economics: They are paid a percentage of the advertising money they spend. The scarcer the ad time, the more it costs; the more it costs, the more agencies spend; the more they spend, the more they earn. That is not a virtuous circle. It's a deathtrap.

Advertising's absurd mass-media economics have spilled over to online. I shrieked when Advertising Age reported that agencies were complaining of a shortage of ad inventory on the home pages of portals including Yahoo. The agencies were creating a false scarcity. There is no end of unsold ad inventory on billions of pages all over the internet. Many of those pages are far better targeted to their needs and would be cheaper and more efficient than Yahoo's home page. Besides, it's not as if a given advertiser's message is going to be seen by everyone who comes to Yahoo, as not everyone goes to its home page. In print and broadcast, advertisers pay for the entire audience—everyone who reads a magazine is presumed to see every ad. Online, advertisers pay only for the pages on which their ads appear—or, with Google's AdSense, they pay only when a reader clicks on an ad. The internet is a more economical and measurable advertising medium but its efficiency is not in agencies' interest because, remember, the more they spend, the more they earn.

Is there any scarcity left in media? Some argue that our attention is shrinking, but I don't buy that. My attention is constant—I have 24 hours in a day, 18 of them awake and 17 of those sober. I choose what to pay attention to in those hours. I believe my attention is more efficient and spent on greater quality than ever, now that I have more choice and more control over my time. Some argue that trust is scarce. Well, I suppose that's always true, but I now have more sources for news than I have ever had—not just my local newspaper but The Washington Post, the Guardian, the BBC, bloggers I respect, and more. Is quality still scarce? Yes, of course, but the more content that is made, the more opportunities there are for more people to make good content. The challenge is sifting through it all to find that good stuff. Where we see challenges, Googlethink teaches us to look for opportunities. Businesses can be built on the need for sifting: commerce sites that find the best merchandise, news sites that read so we don't have to, and entertainment services that gather the critical opinions of the crowd. The internet kills scarcity and creates opportunities in abundance.

Google has found a business model based on creating, exploiting, and managing abundance: The more content there is for it to organize and the more places there are for it to place its ads, the better. If your business is built on scarcity—and most are—you need to ask how you, too, can manage and exploit abundance.

Join the open-source, gift economy

Many a mogul has marveled at the wonder of the open-source economy. The story is often told: Distributed armies of programmers created the most important software underlying the internet, from the Linux operating system that powers most internet servers to the free Apache web server software that delivers most web pages to the 500 million open-source Firefox browsers that show those pages.

Why do these programmers do this work for free? Because they're generous. They want to be part of something. They care. They may want to stick it to the man (namely, Mr. Gates). And they know that banding together in an open network lets them create a better product than they could if they were to work inside most corporations.

How is open-source not chaos? New York University journalism professor Jay Rosen studied the Firefox project when he wanted to bring similar collaboration to journalism at his NewAssignment.net project. He learned that contrary to common misperception, open-source projects are not anarchies. They have leadership and structure. They have people to wrangle the people who want to help. It is elegant organization at work.

Open-source Wikipedia is an incredible resource, a collection of human knowledge vaster and more responsive to change than any encyclopedia. No one who creates it is paid. They contribute out of generosity and ego and because they believe they own it. Note that to make the gift economy work, a project doesn't need its entire community to contribute. Only about 1 percent of those who use Wikipedia create Wikipedia—that is Wikipedia's 1 percent rule. Indeed, if that were doubled, it probably would create chaos. In *Here Comes Everybody*, NYU professor Clay Shirky, who studies social software, calculated the output of the authors of one article: "[O]f the 129 contributors on the subject of asphalt, a hundred of them contributed only one edit each, while the half-dozen most active editors contributed nearly fifty edits among them, almost a quarter of the total." The most active contributor was 10 times more active than the least active.

Wikipedia is not-for-profit. It has spawned a for-profit search service called Wikia, where users are creating even the algorithms that power it. It has commercial competitors, such as Mahalo, a human-powered search and guide created by serial entrepreneur Jason Calacanis, who pays his writers. At the 2008 Burda DLD conference in Munich, Calacanis tweaked Jimmy Wales, founder of Wikipedia and Wikia, for not paying for content. Wales responded that nobody works for free. "What people do for free is have fun. . . . We don't look at basketball games and people playing on the weekends and say these people are really suckers doing this for free." People will contribute their intelligence and time if they know they can build something, have influence, gain control, help a fellow customer (more than a company), and claim ownership.

Customers are also generous with ideas. In 2008 Starbucks launched MyStarbucksIdea.com, where its customers were invited to tell the company what to do (following Dell's lead with IdeaStorm; both use Salesforce

.com's Ideas platform). The response from customers was immediate and impressive: thousands of ideas, votes, and comments. One customer wanted Starbucks to make ice cubes out of coffee so, when they melt, they would not dilute cold drinks; 7,600 fellow customers immediately agreed. Another customer proposed putting a shelf in bathrooms—for where else can you put your drink when you've drunk too much? A few customers found the thought somehow distasteful, but Starbucks called the suggestion a "sleeper idea" that deserved attention.

Some threads emerged from the suggestions and discussion. Many customers wanted express lines for brewed-coffee orders so they could avoid waiting behind alleged coffee aficionados with their half-this, half-that, skinny, three-pump, no-foam, Frappuwhatevers. Some customers asked to be allowed to send in their orders via iPhone. And some customers suggested—and thousands more agreed—that the chain should enable them to program their regular order into their Starbucks card so they could swipe it as they enter, placing the order and paying for it at the same time, letting them skip the cash-register line. One more proposed a pour-it-yourself corner and another asked for a delivery service. The theme—that is, the problem for Starbucks—was clear: long, slow, inefficient, irritating lines. But not one of these customers started with that complaint. Instead, they offered solutions to fix the problem. All Starbucks had to do was ask.

Chris Bruzzo, the Starbucks chief information officer (and a former Amazon executive who learned much about new ways to relate to customers there), built MyStarbucksIdea. The forum was an extension of what Starbucks employees had experienced for years: When they say where they work, "people open up this to-do list in their heads. They have very specific, detailed ideas," Bruzzo told me. Now Starbucks has given them a public platform to share ideas. Because it is open and customers can react to all suggestions, some ideas gain traction and some die on the vine. Customers help the company by eliminating many of the turkeys (such as offering diet powder in drinks or mixing in whole cookies or renaming the accursed Starbucks sizes in honor of the Olympics, from venti to gold). Other ideas take off (such as giving free birthday brews, which Starbucks then considered).

Bruzzo said it is vital for the company to "close that loop in an authentic way and show the commitment on the part of Starbucks to respond to what we've heard, which is about putting those ideas in action or building those ideas together with customers." In short: "We're truly going to adopt it into our business process, into product development, experience development, and store design." To do that, he assigned 48 "idea partners" from all over the company to enter the discussion with customers, using the forum as a laboratory. They were to become champions for ideas back in their departments "so that literally customers would have a seat at the table when product decisions are made." Starbucks, like Dell, has a parallel version of the platform running behind its wall for employees to share and discuss their own suggestions.

Marc Benioff, the outspoken CEO of Salesforce.com, used the Ideas platform first for his own customers and employees and then opened it up to other companies. "It's like a live focus group that never closes," he said in an email. "I believe that these days, the rapid communication that is enabled by wikis, blogs, Twitter, YouTube, and you name it ensures that no matter what kind of company you are, your customers are having a conversation about your products and practices. The question that every company has to ask is, 'Do I want to be part of this conversation? Do I want to learn from it? Am I willing to innovate on the basis of it?' If you harness the power of this community, you will benefit. If you turn your back on it, you get farther and farther out of touch while competitors flourish. The dead-end suggestion box and auto reply are symbols of corporate indifference and are no longer tolerated." (If Benioff sounds like Michael Dell on the topic, there's a reason: He was the one who suggested that Dell needed IdeaStorm.)

Any company or institution could use a platform like this. Governments should use it to gather citizens' suggestions. Editors should use it to solicit and discuss story ideas from readers. Retailers should use it to help decide what goes on shelves. The question is how much companies and institutions are willing to open up to the gift economy and let their constituents take part in their decisions. The gift economy is about more than just listening to customers out of courtesy or respect (now that companies can no longer get away with hiding phone numbers and email addresses and sentencing customers to phone-mail jail). It is about understanding that customers and constituents want to have a voice and gain control. It is a better way to do business. Can customers help design products? Can citizens help write legislation? Can they assign journalists? We will ask those questions in the next section of the book.

Are you willing to have your customers sit at the next desk to work with you? They're willing. Try them.

The mass market is dead—long live the mass of niches

"Masses are other people," sociologist Raymond Williams said in his 1938 book *Culture and Society.* "There are in fact no masses; there are only ways of seeing people as masses."

Advertisers, manufacturers, retailers, media companies, and politicians find it convenient to see us as masses. It's the essence of their business, their efficiency, their reach, their economy of scale. We are their critical mass. So for them, our newfound power to stand out and act as individuals, to coalesce into networks of our own, and to rise above them in Google searches—whether we are competing with them or complaining about them—is a supreme irritant, even a threat.

Mass-based industries and institutions worry now about "fragmentation"—a term used by those who control the mass market. But out here in the market, we call it "choice." Give us more choice and we'll take it. We'll gravitate to our own interests, tastes, and communities. The natural state of life, commerce, and media is choice.

The impending shift away from the mass-market economy was chronicled famously in Chris Anderson's era-defining 2006 book, *The Long Tail.* Anderson said that as the internet creates the means to make, find, and pay attention to an unlimited variety of content about anything, culture and commerce will be less dependent on mass hits. Very few people might watch a single video about how to catch butterflies, but when we can create and watch an unlimited supply of such highly targeted content, the total audience for all these niches together will accumulate to take a sizable share of the audience's attention. In 2008, Anita Elberse challenged Anderson in the Harvard Business Review, arguing that his theory wasn't proving out in practice because a small number of titles still capture a large share of attention and sales. Anderson handily and graciously dealt with her objections on his blog, LongTail.com, reinterpreting some of Elberse's data and definitions to show that the tail, as he measures it, is indeed a factor: Though consumers still buy many copies of a limited number of hit CDs from Wal-Mart, their attention devoted to music *not* sold in Wal-Mart is substantial and growing.

A seminal work in this debate is, believe it or not, a PowerPoint presentation: Umair Haque's New Economics of Media (search Google for "Haque new economics of media" to find it; I also suggest you browse his blog, Bubblegeneration.com, which has influenced Anderson and me). Don't be deterred by his 107 slides and their dizzying economics charts. Haque's lesson is clear: The age of the blockbuster is past. Making money through controlling production, distribution, and marketing is a diminishing game. Haque says media 2.0's three sources of value creation are revelation (finding the good stuff), aggregation (distribution 2.0), and plasticity (enabling content to be extended through, for example, mashups). This economy, he says, requires openness, decentralization, and connectedness through niches—not blockbusters. The new opportunities lie in the long tail.

I know the arguments to the contrary: the Oscars, the Olympics, *Harry Potter*, *The DaVinci Code*, *American Idol*, Wal-Mart. Yes, stipulated, there will still be blockbusters. But let's also agree to these factors: The tools that enable anyone to create and distribute goods and media will yield almost unlimited choice. The public will increasingly seize upon that choice. The attention given to and thus the value of this new wave of choice will grow. There are new opportunities in enabling, organizing, and monetizing this abundance. The blockbuster strategy always was a gamble; as it continues, it is a bigger gamble than ever. The mass market's hold over the economy diminishes.

The mass market was a short-lived phenomenon. It began with the large-scale adoption of television in the mid-1950s—and the consequent death of second and third newspapers in most American cities, yielding one-size-fits-all mass products in both broadcast and print. It was in the mid-1980s, in the age of the remote control, that I became the TV critic at People magazine, the last great mass magazine launched in America. In its first decade, the magazine was pretty much a piece of cake to run: Put a star in a big show on the cover and watch it sell. But I remember the day that ended, when my editor and mentor at People, Pat Ryan, yelled at me from the other end of the hall: "TV's dead, Jarvis! It's dead!"

She had just received the latest in a string of bad sales reports on covers featuring stars in top shows. They didn't produce guaranteed hits anymore because Americans were not all watching the same shows. No longer did we wake up as one nation asking, "Who shot J.R.?" Instead, while I was watching MTV, you were watching the History Channel, she was watching the Golf Channel, and the kids were using that new-fangled VCR with its flashing "12:00" or playing video games (never mind that the internet and the iPod had yet to come). Some lament the death of the allegedly grand shared experience of mass media, portraying it as the electronic fireside around which we sat in a common cultural encounter. I don't. I value choice.

The fragmentation of media threw business strategies into a dither. Advertisers still wanted to buy us en masse, so media had to work harder to find a mass to satisfy them. It was then that People shifted from covering the event in the star's career to the event in the star's life, and other publications followed the lead. Bodily fluids journalism, I called it: stories about celebs' deaths, diseases, affairs, scandals, weddings, babies, divorces. The balance of power at mainstream publications—at least on their covers—shifted from news to celebrity, journalism to gossip, editor to PR person, hack to flack. Stars realized the dollar value that their names and faces brought to magazines, and that's when their publicity people took over. Editors used to act as gatekeepers to the most valuable commodity—the audience. But then PR people became gatekeepers to a more valuable asset—celebrity. They would negotiate access, guarantees of covers, photo approval, selection of reporters, and even the ability to pick and change their quotes. PR people held such power because they now held the key to magazines' ability to attract large audiences. Magazines all wrote about the same celebrities and scandals—they went more mass—but that was because there were fewer topics that would attract lots of buyers. Too many of us were busy watching Discovery instead of *Dynasty*. And that, in turn, changed the economics of TV content. Networks seeing their shrinking masses could less afford to gamble on expensive dramatic shows, let alone miniseries (remember them?). They were replaced by so-called reality TV, which was not only cheaper but also more sensational.

What replaces the mass? The aggregation of the long tail—the mass of niches—does. We each gravitate to our own interests and, thanks to the new and inexpensive tools of content creation online, there's sure to be something for everyone—and if there isn't, we can make it ourselves. The 500-channel world never materialized. Instead, the billion-choice universe emerged. Internet tracking service comScore said in 2008 that we watched

10 billion videos a month online. Of course, none of them individually had the ratings of the Super Bowl. But together, those 10 billion videos captured a huge amount of our attention. eMarketer says 94 million Americans read 22.6 million blogs in 2007—more than there are newspapers and magazines: blogs about knitting, blogs about heart disease, niche blogs about writing niche blogs. As I write this book, Wikipedia has 2.3 million articles and even it has new competitors, including the *Star Wars* version, Wookiepedia. Everyone, and every interest, has a place online.

Advertisers, addicted to one-stop shopping, still spend huge budgets on TV that are way out of proportion to the time the audience spends there versus the time we now spend on the internet. That can't last forever. Soon, agencies will have to work for a living. Instead of reflexively buying slots on primetime TV, they will need to put together networks of smaller media with smaller audiences that add up to a critical mass. This approach is harder but more targeted and more efficient. Why advertise diapers on a show I watch—next to my teenage kids—when instead Pampers can now advertise on mommy blogs?

As advertisers and agencies catch up with the death of the mass market, money will flow online that will, in turn, support the creation of new content, which will draw a greater audience, which will earn more money. On and on this virtuous circle will go until broadcast TV is a shell of its former self. There will still be hit shows—the *Deal or No Deals* that pass for our grand shared experience today—but there will be fewer of them.

Google figured out how to navigate the universe of niches and profit from it. It created a new way for advertisers to reach highly targeted audiences just as they search for and read relevant content. Even more disrupting to old media ways, Google didn't charge for eyeballs—that is, the size of the audience—but for clicks—that is, action. Advertisers could measure the return on their investment instead of talking to faceless masses that may or may not have been listening. More disruption: Google didn't set ad rates as old media did; it let the marketplace set the price of keywords in auctions. Because Google benefits as more ads are clicked on, it is in Google's interest to continue to improve its targeting and effectiveness. That improves both advertisers' efficiency and Google's bottom line. This virtuous circle of virtuous circles is how Google built its empire around the fall of the mass and the rise of the niche.

You, too, must learn how to make the transition from mass to niche and how to exploit it. If you're still selling products to the masses, you're going to find it harder. If you're making one-size-fits-all products, realize they don't fit everyone. Customers will tell you what they want instead. In the next section of the book, we will examine scenarios for adapting and capitalizing on the move from mass: how automakers should let us help them make cars, how retailers can help us find unique goods, how universities should help us craft our educations. The shift from mass is really a shift of power from top to bottom, center to edge, them to us.

The mass market is dead. It committed suicide. Google just handed it the gun.

Google commodifies everything

In the earliest days of the web, I watched focus groups where users thought there was this amazing new company that had acquired all the content you could imagine about every subject possible, as if from the merger of a library, a newspaper, a magazine, and a weather service. That company was Netscape. It merely made the first commercial browser that took readers to those other companies' sites. But Netscape got the credit.

Today, that amazing brand is Google. People go online looking for something, find the answer, and often don't know where they found it. Google found it. They're savvier today and know that Google doesn't own all the content it links to. But that doesn't matter, so long as they find what they want—and Google is damned good at that. That's great for users but bad for brands. Here you work your buns off creating a brand online; you build technology and staff to maintain your site; you spend a fortune on marketing and search-engine optimization to get people to find it; you tell advertisers how many users come to your page and like your brand. But in the end, huge numbers of users don't recall coming to your site and don't credit your brand. When I worked on newspaper sites, we knew we had more users than the research said. The problem was, when users were asked where they had seen a piece of information that could have come only from us, they often couldn't remember. Google found it for them. Google diluted our brands.

Google has turned commodification into a business strategy. Content is commodified: Google makes it just about as easy for you to find what I've written on a topic as what Newsweek has written. Once was, brands organized information but now Google does. Media are commodified: Google places marketers' ads on sites without telling them where the ads will appear. It places those ads not as an ad agency would—on the basis of the audience size, demographics, trust, or value of a media brand—but on the coincidence of words on a page. The value of the ad depends only on how many people click on it. Thus the media brand behind the content where the ad appears becomes less critical and less valuable. Even the audience is commodified: There's little that distinguishes one of us from another—not age, income, gender, education, interest, all the things advertisers historically paid for. Everybody's like everybody else. We're just users. We might as well be pork bellies. And advertisers are commodified: Their text ads look alike, without their expensive logos and brand messages. You'd think they'd object, but they don't mind because they pay only for clicks. Google has cleverly reduced the risk in advertising, so advertisers let Google drive.

It isn't all bad. The leveling of the playing field the internet and Google engineered also made it possible for a tiny store selling a niche product to find its ideal customer or for a mere blogger to swim alongside big, old media. But in that process, it's ironic that our unique identities, personalities, brands, qualifications, interests, relationships, and reputations as publishers—the values the internet enables—can be lost even as we can be found via Google.

What do we do about the threat of commodification? One smart response is to play by Google's rules and take Google's money as About.com has done. Or you can join networks with other specialized niches to reach critical mass, as Glam.com has done. Or get people to link to you and talk about you because you're just so damned good, as Apple does. Or place your ads on highly targeted sites where you know your customers are, sponsoring that mommy blog with free baby food for loyal readers. Develop a deep relationship with your constituents so they come back to you directly, not just through Google search but by using social services such as Facebook. Serve the niche well rather than the mass badly.

Welcome to the Google economy

In April 2008, just as America was diving into recession, Google announced another amazing and profitable quarter. The New York Times story was headlined, "Google defies economy." It should have read, "Google defines economy."

Old definitions of our economy measured the performance of big companies and their impact on each other (see: the Dow Jones Industrial Average). Media and advertising served only large companies because only they could afford to advertise in large outlets. Manufacturers could get retail space only if they operated under the economies of scale. That was the mass economy. Then Google's marketplace for advertisers of all sizes introduced the small-is-the-new-big ecosystem, the mass-of-niches economy.

That Google's advertising is run in an auction marketplace means that its economy is more fluid; it fills in voids. When an economic downturn affects, say, travel, a magazine such as Condé Nast Traveler will suffer—airlines and resorts will advertise less and there aren't more big advertisers to fill the gap at Traveler's price. But on Google, if American Airlines and the Ritz aren't buying the keyword "Paris" this month, other advertisers may buy it. The price of that keyword may decline with demand, but in Google's very broad economy, the prices of other keywords (e.g., "foreclosure" and "credit") may rise.

This practically unlimited supply of advertisers in a fluid marketplace appears to be a new economic model that may insulate Google from some of the dynamics of an economy built on mass and scarcity. Google has its own economy.

Google also reflects our new and emerging economic reality. In the financial meltdown that reached full flame in the fall of 2008, we saw not just the failure of mortgages, derivatives, banks, and regulation. We saw the dawn of a new economy that could best be viewed and understood through the lens of Google, the one company that—by design or by luck—was built for the emerging world order. As banks, companies, and even nations faltered, Google still announced profits rising 26 percent.

In Google's economy, companies will no longer grow to critical mass by borrowing massive capital to make massive acquisitions—at least not for the foreseeable future. Instead, they need to learn from Google and grow by building platforms to help others prosper. Indeed, growth will come less from owning assets inside one company and amassing risk there than from enabling others in a network to build their own value, reducing their cost, and spreading their risk. That is Google's way.

New Business Reality

Atoms are a drag
Middlemen are doomed
Free is a business model
Decide what business you're in

Atoms are a drag

Stuff is just so last-century. Nobody wants to handle stuff anymore. It's inconvenient and expensive. If you have stuff, to paraphrase the late, great George Carlin, you have to find a place for it. You must buy the raw stuff you'll use to make your stuff. Then you have to store your stuff, pack it inside more stuff, and ship it along with other stuff. Not to mention that you have to pay to hold an inventory of stuff and you risk your stuff going out of style, in which case you'll be stuck with a lot of useless, old stuff. Anybody can reverse-engineer your stuff and make the same stuff. Now you may argue that their stuff isn't as good as your stuff—as Carlin said, "Have you noticed that their stuff is shit and your shit is stuff?"—but once there's competitive stuff, you'll probably have to charge less for your stuff to sell more of it. Stuff is a pain. Digits aren't.

Since the dawn of industry, controlling things and the means to make, market, and distribute them has defined businesses. Carmakers sold cars, newspapers papers, book publishers books. They identified—and limited themselves—by their products. We are what we make.

Magazine companies sell what? Magazines? Not so fast. In 2008, Colin Crawford, an executive at the tech publisher IDG Communications, bragged that his was no longer a print company. IDG had crossed the Rubicon from print to digital two years earlier when its growth in online revenue exceeded the decline in its print revenue. As a result, Crawford blogged, his team could focus on "the changing needs of their customers" and on new online and mobile products and event businesses. The staff was, he said, "unburdened by print."

Print had become a burden to a print company. It's expensive to produce content for print, expensive to manufacture, and expensive to deliver. Print limits your space and your ability to give readers all they want. It restricts your timing and ability to keep readers up-to-the-minute. Print is already stale when it's fresh. It is one-size-fits-all and can't be adapted to the needs of each customer. It comes with no ability to click for more. It can't be searched or forwarded. It has no archive. It kills trees. It uses energy. And you really should recycle it, though that's a pain. Print sucks. Stuff sucks.

So who wants stuff? Not Amazon. Yes, Jeff Bezos built a great company around selling things: books, gadgets, hardware, almost anything that can be delivered to our door. Just as Craig Newmark of craigslist is blamed (unfairly) for driving a stake through the heart of papers, Bezos is blamed for crippling bookstores, with independent outlets dying and even chains suffering. But who can blame shoppers for going to Amazon with its discounts, convenience, and selection?

Bezos is as clever about stuff as he can be. He holds as little inventory as possible, getting more merchandise as needed when we order it. He owns no stores, pays no retail rent, hires no sales clerks. He doesn't own the shipping infrastructure he uses but gets the best possible deal he can with outside services because he wields such huge volume. Because of that volume, he negotiates the best prices from suppliers. He passes a portion of those savings—the internet dividend—to his customers, which only builds his volume even higher. It's a business of efficiencies, volume, turnover, and tight margins.

I bought Amazon stock and I'm holding onto it, not because Bezos built the better bookstore but because he is creating digital equity. He sells his retail services to other merchants, sending them customers online and taking a cut, in some cases warehousing and shipping their inventory and charging for the services. He also took the computer infrastructure he had to build and offered it to any company as a low-cost, pay-as-you-go service: computing power, storage, databases, and a mechanism for paying programmers. Countless companies now use Amazon Web Services as their backend, foregoing or at least forestalling investments in computers and software. Amazon has also created the infrastructure for an on-demand workforce called Mechanical Turk (named after a phony chess-playing automaton from 1769 that had a human chess master hidden inside). Companies post a repetitive task to be done and anyone can earn money— as little as one cent per task—by verifying the address in a picture, for example, or categorizing content. It's a flexible marketplace for labor. With all these services, Amazon is supporting a wave of entrepreneurial effort. Why would a bookstore do that? Amazon turned its cost center into a profit center—and beat Google to the opportunity (Google later followed suit).

Bezos is not building a stuff company. I believe he is building a knowl-edge company. No one knows more about what identifiable individuals buy than Amazon—not even Wal-Mart (to them, we are mostly a mass) or the credit card company (they can't necessarily see what products we buy at the grocery store). Amazon knows what we bought, when we bought it, and what else we bought with it. It can try out sales pitches to see which work best. It knows enough to predict what we might want so it can en-tice us to buy it. It has captured millions of reviews and ratings of every imaginable product from people who have bought and used them: a more valuable repository of consumer reports, I'd say, than Consumer Reports itself. No one knows more about the stuff we buy than Bezos. Handling stuff becomes a small price to pay to become so smart.

Amazon is positioned perfectly for the transition to digital content de-livery. It is selling and delivering books to PCs and its Kindle e-book reader. It is selling movies direct to our TV sets. It is selling music down-loads. Amazon has built a strong position in content thanks to innovations ranging from reviews to searching inside books to automated recommen-dations. By reflex, many of us go to Amazon to check out products before we buy them. That is Amazon's brand and value, as much as the stuff it sells.

Bezos built a digital knowledge and service empire. Just as fast-food joints make more money selling Coke than cheeseburgers and some retail chains have built more value in real estate than merchandise sales, Bezos doesn't really make his money pushing atoms. Like Google, he creates value by getting smart and building bits.

Are you limited by your stuff? If a magazine publisher no longer thinks of itself as a magazine company and if a bookstore can build a knowledge company, then ask what you can be. Where is your true value? I'll bet it's not in the atoms you move around. It's in what you know or how you serve or how you can anticipate needs, isn't it?

Middlemen are doomed

Nobody likes a middleman. Well, except for my very nice literary agent. When she read in the proposal for this book that middlemen were doomed, she protested: "But that's me, Jeff." Sorry to say, yes. When she sold the book to the publisher, you could say that she sold her own professional obit.

Then again, she did make the sale. Without her—and her relationship with publishers—my proposal wouldn't have gotten into three houses, which led to an auction that raised the price (boy, was that fun). Even though my agent charges a higher commission than a real-estate agent (much higher), she increased my advance by more than the amount of the commission she was paid. Her agency also provides editorial, legal, and marketing advice. My agent made the marketplace more efficient and added value for me. She also makes the business more efficient for publishers, sifting through an abundance of book ideas and writers.

When I went to work as an online executive at a media conglomerate in the 1990s, I was delighted that I would get to work with a book publisher as it went online. But my boss warned me that I shouldn't get too excited. He explained that a publisher doesn't have direct relationships with readers (bookstores do) or even with talent (agents do). Publishing, he said, is a distribution business. Publishers, too, are middlemen.

Today, technology and the internet have fostered new self-publishing companies—Lulu.com, Blurb.com—that enable authors to have their books designed, printed, sold, and distributed and to keep a much higher proportion of the sale price—up to 80 percent, versus the roughly 15 percent of the hardcover price authors receive from mainstream publishers (minus the agent's 15 percent). Authors can sell their books directly to readers via Amazon as well. But, of course, mainstream publishers will argue that because they have relationships with bookstores for sales and with media outlets for promotion, they are able to sell many times more books for a higher profit than an author can when selling directly. They'd be right—for now. This is why even I, cyberguy, chose to have this book published the old-fashioned way: Because the book and my ideas will be distributed and promoted broadly and I will likely make more money. My publisher is adding value. In the next section of the book, we examine how book publishers need to update their business and their books for the Google age.

For all middlemen, the clock is ticking and the question of value is looming. Every time Google makes a direct connection, a middleman's value is diminished. Are you a middleman? If the web is hurting rather than helping your business, the answer is probably yes. If you make the marketplace more efficient, if you solve problems of abundance and confusion and add value, good. But even if you do, anyone can use the internet to undercut you—to craigslist you. If you make your living telling people what they can't do because you control resources or relationships, if you work in a closed marketplace where information and choice are controlled and value is obscured, then your days are numbered. I'm talking to you, car salesmen, advertising agencies, government bureaucrats, insurance-office benefit-deniers, head hunters, travel agents (oh, sorry, they're already nearly extinct), and real-estate agents.

The internet abhors inefficiency, eliminating it whenever Google, Amazon, eBay, craigslist, et al connect buyer to seller, demand to fulfillment, question to answer, SWF to SWM. Economist Umair Haque, blogging for the Harvard Business Review, sees a shift from an economy built on inefficient marketplaces, where ownership and control are centralized, to an economy built on efficiency, where information is open and the power resides nearer the edges. "Competitive advantage is fundamentally about making markets work less efficiently," he said. "One catastrophically effective way to do that is to hide and obscure information—to gain bargaining power relative to the guy on the other side of the table." The new way to succeed is to do the opposite: "Release information bottlenecks and make things more liquid." In other words, stop trying to make money by interfering in transactions.

Consider my least favorite inefficient marketplace: real estate. I hate paying agents 6 percent commission for doing so little. They, in turn, hate it when I talk about them on my blog. What we think of real-estate agents around the world is an open secret. A 2008 survey by the British Journalism Review found that real-estate agents are the least-trusted professionals, worse even than tabloid reporters. Only 10 percent of Britons trust them.

But real-estate agents have nothing to fear from me—or, they think, the internet—because they control one of the last dark pools of restricted information left in business: the multiple-listing service (MLS) database of properties for sale. If your house isn't listed there, buyers won't see it and other agents won't show and sell it. But only real-estate agents can list homes in the MLS. I call that monopolistic restraint of trade. Real-estate agents call it service. The U.S. Justice Department called it antitrust, and in 2008 it reached a settlement with the National Association of Realtors to open up the multiple-listing service somewhat to discount brokers. It was a small victory against the middleman.

Agents say they bring you pricing expertise. But in the U.S., Zillow. com will give you an automated estimate of your home's worth based on comparable sales in your area. Zillow tracks its own accuracy, comparing actual sale prices with its estimates. So much for that bit of expertise from Ms. Agent.

Agents say they market your home. Pshaw. They used to advertise a selection of homes in the Sunday paper, but those ads promoted their agencies as much as marketing specific properties. Real-estate ads are like grocery ads that entice you to come in because flank steak is on sale or because one house caught your eye. Now, thanks to the internet, there's less need for agencies to advertise in papers. Real-estate agents can save money by putting listings on their own web sites or even on craigslist and Zillow. They rarely pass those savings on to homeowners.

Agents say they bring their expertise to buyers, not just sellers. When I bought homes, I went to agents so I could see the multiple-listing service and pick out my prospects. The only real service the agent provided was hauling me around and letting me into homes.

"A real-estate agent may see you not so much as an ally but as a mark," Steven D. Levitt and Stephen J. Dubner wrote in their 2005 paean to seeing things differently, *Freakonomics*. They cited a study that found that real-estate agents keep their own homes on the market an average of 10 days longer than homes they represent—and agents sell their own homes at prices 3 percent higher. Levitt and Dubner explained that it's more efficient for agents if they can get you to sell quickly, even if for a few dollars less. "Here," they wrote, "is the agent's main weapon: the conversion of information into fear." In the long run, Zillow and similar services will become smarter than the smartest agent. On the internet, more information equals more power and value. (In the next section of the book, I'll outline how I propose to replace real-estate agents.)

In the early 1990s, when I worked with newspapers, I predicted that real-estate agents would desert papers for online. I advised newspapers to get into the real-estate business themselves, becoming agents just so they could get access to the multiple-listing data. Newspapers are not in the publishing business but are in the information business, and the MLS is the key to the information that mattered. God no, the publishers said, we don't want to rock the boat with agents and lose that ad revenue. But papers were bound to lose it anyway.

Newspapers didn't know what business they were in and didn't know who their true customers were. They thought they were in the business of selling real-estate ads, not serving homeowners (aka readers). Newspapers even tried to discourage homeowners from posting their own for-sale-by-owner ads because agents saw those homeowners as competition. Staying loyal to real-estate agents over readers did newspapers no good. The agents didn't return the loyalty. Newspaper classified revenue in real estate, jobs, and cars fell from $19.6 billion in 2000 to $14.2 billion in 2007 (adjusted for inflation, that's a drop of about 40 percent). If newspapers had seen just how dire their future was, they might have gone around the agents they had protected and freed up information for readers. But soon it was too late. Though real-estate ads increased as home prices rose, the home bubble eventually burst in 2008, and papers' last gravy train derailed.

Real-estate agents and papers are not alone as middlemen, the proprietors of inefficient marketplaces. The monopolies, duopolies, oligopolies, cartels, and controlled marketplaces enjoyed by cable companies, phone companies, broadcasters, advertising agencies, health-care companies, and government are challenged by the internet's open marketplace of information. Google isn't their competitor. Google is the weapon their competitors wield.

Free is a business model

Free is impossible to compete against. The most efficient marketplace is a free marketplace. Money gets in the way. It costs money to market and to acquire customers so you can sell things to them. It costs money to take payments. Charging customers stops some unknown number of them from getting your product or using your service, which stops you from having a relationship with them. Money costs money.

Obviously, that's absurd. The goal of any business is to collect money and make a profit. The most sensible way to do that is to charge the people who consume what you produce, right? Not always. Return to the chapter on networks, where new-age phone companies (Skype), retail marketplaces (Amazon, eBay), and classified-ad marketplaces (craigslist) grow larger by charging less, even nothing.

As much as I abuse media for operating under the rules of the old economy, let it be said that more than a century ago they created a new model built on not charging customers full freight. Rather than making readers or viewers cover costs, media charge the people who want to reach the people they reach—they charge advertisers. That is what makes broadcast free and newspapers and magazines inexpensive. A high-end magazine might cost $4–$5 per copy to produce and distribute; it might cost another $20–$30 to market to acquire that subscriber. Yet many successful monthly magazines charge their readers only $1 per copy to subscribe; it's nearly free. A publisher in this scenario is in the hole $50 or more per subscriber in the first year (that improves every time readers renew). Clearly, though, magazines make money from ads—enough to dig themselves out of that hole and earn an impressive profit through the side door.

Google and the internet have created more models for making money through that side door. The appeal of this path is that often you need not own the assets that make you money. Google doesn't want to own the content it searches; it wants knowledge to be free online so it can organize more of it. In the late 1990s, Google executives came to me when I worked for a magazine publisher, trying to convince us that we should take all our content archives—for which we charged readers—and put them on the internet for free. Google search, in turn, would send lots of traffic to the old articles. Google also offered to put its ads on these pages, making new money on old content—more money, they assured us, than we were making from archive fees. They were probably right, but I knew it would have been impossible to convince magazine publishers—who were too accustomed to owning and exploiting their valuable assets—to see value elsewhere. At the time, publishers didn't understand that restricting access online was turning away people to whom they could show ads and sell magazines and build relationships. The pay wall was less a revenue opportunity than an opportunity cost.

The New York Times learned this lesson through experience. So accustomed were Times executives to selling papers and charging readers for access to content that they couldn't bear the idea of giving it all away on the free web. They decided to charge readers online and had to find something to put behind a pay wall. It had to be something that would not hurt their advertising business (they wouldn't have wanted to put ad-rich travel content behind a wall, losing audience and ad revenue, for example) but something that readers thought was still worth paying for. In 2005, NYTimes.com fenced off columnists and archives along with other goodies and charged $49.95 per year for access. TimesSelect got 227,000 paying customers (plus print subscribers and students, who received it for free). It brought in a reported $10 million annual revenue. I never saw an accounting of the cost of marketing to acquire those subscribers or of customer service; the profit margin was not reported. In a speech then, Guardian editor Alan Rusbridger showed a picture of The Times' lavish new headquarters and said that revenue wouldn't pay the gas bill in the building.

The Times killed the service in 2007 and freed its content again for a few simple reasons: First, it increased the audience to the paper's site; within months after tearing down the wall, audience increased, by one account, 40 percent. Second, The Times could make more money on the advertising shown to its additional audience. Third, opening up improved the paper's Googlejuice by bringing in more clicks and links, which in turn yielded more traffic. Finally, the dropping of the toll booth brought The Times' columnists back into the conversation. Rusbridger had said The Times walling off its columnists was "brilliant" for the Guardian because it opened the door for it to reach The Times' former readers (the Guardian gets a third of its audience in the U.S.). In the end, The Times rediscovered the value of free.

Google understands the value of free better than anyone. When it bought Blogger, it stopped charging for the service and added advertising. When it launched Gmail with tons of storage, it made the service free and served targeted advertising. More recently Google has set out to pull a craigslist on the $7 billion mobile directory-service business. Google made directory assistance free at 1-800-GOOG–411. My accursed mobile service provider still charges me $1.79 to find a number—and mind you, the only reason I'm looking for a number is so I can make a call using the company's network, which I pay to do. This is like a store charging us for directions to come spend money there.

Google surely will make money on its mobile directory service with advertising. It will learn more about our behavior and needs. I can imagine it using us to create a vast repository of our reviews and recommendations about establishments ("leave your review after the tone" or "rate the restaurant using your keypad"). Google may find yet another side door to make money. Tech publisher Tim O'Reilly speculated on his blog that Google wants to gather billions of voice samples as we ask for listings. That will make its speech recognition smarter, helping it get ready for the day when phones and computers respond to voice commands. Chris Anderson, editor of Wired magazine, projected that by 2012, Google could make $144 million in fees from users if it charged for directory assistance, but by foregoing that revenue it could instead make $2.5 billion in the voice-powered mobile search market. As with newspaper classifieds, the entire industry may shrink but the winner will grab the biggest share of what is left. That winner is likely to be a new player, not one trying to protect old revenue streams and assets. By making its service free, Google will establish itself as the leader in providing local information and position the company for the coming mobile explosion. On Jim Cramer's CNBC show, Google chief Eric Schmidt said the company anticipates making more money on mobile than desktops because mobile provides a better way to targets ads, and targeting is Google's real strength.

Anderson, author of *The Long Tail,* argues in his next book that free is a business model. In a preview of *FREE!* in Wired, he provided a case study: Ryanair, a discount flier out of Dublin, has been selling tickets around Europe for as little as $20 and hopes to offer seats for free. The airline saves money—and who can complain at these prices?—by using less popular airports. Once it has you, it charges extra fees for priority boarding, luggage, food, and credit card handling (American airlines have started similar charges but at higher ticket prices and with spotty service). Ryanair also shows ads onboard—an ideal exploitation of a captive audience. It hopes to start onboard gambling, which could be a huge money-maker.

A favorite buzz phrase of consultants in the last few decades is "zero-based budgeting"—rethinking and rebuilding your business from scratch, without legacy structures and assumptions. Now you really can start at zero: What if your goods cost you nothing? What if you charged nothing? Where does your value exist? What is the essence of your business? What can you learn from it? How do you make money—is there a side door? Your business will likely operate at a different scale: It could be smaller but with lower costs and higher margins. Or it could be larger with lower margins, which helps it grow bigger faster with less investment and risk. But it will surely be different. Rich Barton, founder of online real-estate service Zillow, told The New York Times: "The internet is a great big race to free. Anyone who has built a business model with a price above free for something that can be free is in a tough strategic position."

So how do you get to free first?

Decide what business you're in

What business is Google really in? Of course, it's in the search business; that's why we go there. But it doesn't profit from licensing its search. It is also in the service business, providing us with everything from email to document management to mapping to publishing tools to social networks to telephone directory assistance to video distribution. But it charges us for none of that. It is not in the stuff business, moving things or selling them (though it has not fully escaped the tyranny of matter; it buys a lot of atoms in the form of computers, and it has to spend a lot on charged atoms to power them). It is also not in the content business; apart from its collaborative, Wikipedia-like Knol, it doesn't create or control original content but instead prefers to organize others' (owning content would put Google in competition with the businesses whose content it exploits). Ultimately, Google is in the organization and knowledge businesses. Google knows more about what we know and want to know and what we do with that than any other institution. But its profit doesn't come from that either. Google's profit comes from advertising, which it dominates because it is so good at search and has so many of us using its services and knows so much that it can target ads efficiently. Google knows what it is.

AOL thought it was in the content business, which is why Time Warner, a content company, made the disastrous mistake of combining with AOL. In reality, AOL was in the community business (its chats and forums were pioneering and popular, long before Facebook or MySpace) and the service business ("you've got mail" on AOL way before you've gotten it from Gmail). AOL didn't ask the right question: What business are we *really* in?

Poor Yahoo thought it, too, was in the content business; that is why it hired a Hollywood studio exec, Terry Semel, to be its CEO. He tried to turn it into a digital movie studio. But Yahoo could have owned search as the pioneer in web directories; it handed that to Google. It could have owned search advertising as a pioneer there, too, but it ceded leadership in automated advertising as well to Google. What business is Yahoo *really* in? I think it never decided.

What business are you *really* in?

Many companies worry that they can't make the transition to bits: analog to digital, physical to virtual, 1.0 to 2.0. Some are nearer than they think. Kodak is a classic case of a company said to be making the transition from atoms to bits—physical film to digital images, sales to service. If it had realized soon enough that it was in the image and memories business—if it hadn't defined itself by the atoms it pushed and processed—it should have beaten Yahoo to the punch and bought the photo and community service Flickr. When I think of pictures today, the first brand that comes to my mind is Flickr. Others think of Google's Picassa. I also think of my Nokia camera phone. Who now thinks of Kodak (or Polaroid, which stopped making instant film cameras in 2008)? No one.

Airlines are the ultimate atomic enterprises, moving our own molecules from place to place and burning lots more molecules in the process. But even airlines could be relationship and knowledge companies. Are cable companies pipe managers, or should they be hosts for our digital creations? Are doctors' offices sickness companies or health companies? Are insurance companies arbitrageurs of risk or guarantors of safety? Are grocery stores stuff companies or knowledge factories? Are restaurants kitchens or communities? We'll examine such upside-down views of these industries and more in the next section of this book.

You should be asking: Am I a knowledge company? A data company? A community company? A platform? A network? Where is your value and where is your revenue? Remember that they might not be in the same place; the money may come in through a side door.

It's time for your identity crisis.

New Attitude

There is an inverse relationship between control and trust
Trust the people
Listen

There is an inverse relationship between control and trust

Trust is more of a two-way exchange than most people—especially those in power—realize. Leaders in government, news media, corporations, and universities think they and their institutions can own trust when, of course, trust is given to them. Trust is earned with difficulty and lost with ease. When those institutions treat constituents like masses of fools, children, miscreants, or prisoners—when they simply don't listen—it's unlikely they will engender warm feelings of mutual respect. Trust is an act of opening up; it's a mutual relationship of transparency and sharing. The more ways you find to reveal yourself and listen to others, the more you will build trust, which is your brand.

Give the people control and we will use it, my first law decrees. Don't and you will lose us. In a meeting of web 2.0 gurus at National Public Radio sometime ago, I heard David Weinberger—coauthor of *The Cluetrain Manifesto*, author of *Everything's Miscellaneous*, and a Harvard fellow—extend that law. He may have thought of this law as his own, but I prefer to co-opt it as Weinberger's Corollary to Jarvis' First Law: "There is an inverse relationship between control and trust." There's another one of those counterintuitive lessons of the Google age: The more you control, the less you will be trusted; the more you hand over control, the more trust you will earn. That's the antithesis of how companies and institutions operated in pre-internet history. They believed their control engendered our trust.

In the early days of the internet, some journalists dismissed new sources of information—weblogs, Wikipedia, and online discussions—arguing that because they were not produced by fellow professionals, they could not be trusted. But the tragic truth is that the public does not trust journalists. A 2008 Harris survey found that 54 percent of Americans do not trust news media, and a Sacred Heart University poll said that only 19.6 percent believe all or most news media. In the U.K., a 2008 YouGov poll found what looks like a high number who trust BBC journalists a great deal or a fair amount—61 percent—but that was down 20 points since 2003.

Trust is—no surprise—an issue with political leaders. In 2007 the World Economic Forum released a Gallup Voice of the People survey reporting that globally, 43 percent of citizens said political leaders are dishonest; 37 percent said they have too much power; 27 percent said they are not competent. Fifty-two percent of U.S. citizens said their politicians are dishonest. Business came off only marginally better: 34 percent believed business leaders are dishonest; 34 percent said they have too much power.

To co-opt Sally Field: We don't like you. We really don't like you.

When asked how to restore trust, a plurality of world citizens polled by Gallup—32 percent—argued for transparency and 13 percent pushed for dialogue with consumers. There is Weinberger's Corollary in action: Open up, hand over control, and you will begin to regain the trust you have lost.

Trust the people

Before the public can learn to trust the powerful, the powerful must learn to trust the public.

I learned my lesson about trusting the people when I was a TV critic at People magazine in the mid-1980s. That was the critical moment in the history of popular culture when the remote control passed 50 percent penetration on American couches. The remote, the cable box, and the VCR reached critical mass, and together they put us in control of our consumption of media. No longer were we imprisoned on *Gilligan's Island* by the bad taste of network programmers in Burbank.

At the end of a season back then, I was about to go on a CBS morning show to talk about the season's ratings when the segment producer, Bonnie Arnold, came to me on the set and summarized what she thought I would say: "You're defending the taste of the American people, right?" I recoiled in horror. I'd never do that, I said. How could I say that the masses have taste? I'm a critic, a media snob. This is *television* we're talking about. Vast wasteland, remember? Arnold argued with me: "You're saying that good shows rose to the top of the ratings and bad shows fell. So you're arguing that the audience has good taste."

Ding. She was right. I *was* defending the taste of the American people. That moment evolved my worldview (as the internet would again 20 years later). I realized just then that once the people were given choice and control, they would tend to pick the good stuff. The more choice they had, the better the stuff they picked. The better stuff they picked, the more Hollywood was forced to make good shows for them. Here was a virtuous cultural circle and another law: Abundance breeds quality.

Of course, there have been exceptions—blooper shows, game shows, tabloid shows, trailer-trash talk shows. We have feared that each of these trends would take over television and society. But in each case, we as a nation overdosed on our guilty pleasures and they faded away. Quality wins. I've long argued that the golden age of television was not the 1950s, with our misplaced nostalgia for its cheesy video Vaudeville. Uncle Miltie, I say, was a hack. *The Sopranos* is higher art than *Playhouse 90*. *Seinfeld*, *Cheers*, and *The Office* are funnier than *The Honeymooners*. Sacrilege, perhaps, but true. The golden age of TV is now—or probably tomorrow, as TV is reinvented and opened up on the internet.

On that day in the 1980s, I learned to trust the people. Bonnie Arnold's challenge turned me into a populist. I realized that if you didn't trust the people, then you couldn't believe in democracy (why let us pick our leaders . . . even if we sometimes do bollix it up?), free markets (shouldn't somebody be in charge?), journalism and education (why inform the people if they're a bunch of idiots?), even reform religion (surely the masses shouldn't talk directly with God).

My new, populist worldview was only strengthened by my experience with the internet, which gives us control over not just our consumption of media but now its creation. The internet enables unlimited creation and, because abundance breeds quality, we now have more good stuff. I know, you're going to rub my nose in a YouTube video—one featuring a flaming fart or a twirling cat—and you'll argue that the internet opens the door to the creation of crap. That, it does. But it also offers new opportunities for talent and new stages for voices that could not emerge in the old systems of control. There have always been bad books on bookstore shelves next to the gems. See: Danielle Steel. There will always be flaming cat videos next to art online. But there is the opportunity to make more art now. The challenge is finding and supporting it. That is where Google comes in. Google can't and shouldn't do it all; we still need curators, editors, teachers—and ad salespeople—to find and nurture the best. But Google provides the infrastructure for a culture of choice.

Google's algorithms and its business model work because Google trusts us. That was the *ding* moment that led Sergey Brin and Larry Page to found their company: the realization that by tracking what we click on and link to, we would lead them to the good stuff and they, in turn, could lead others to it. "Good," of course, is too relative and loaded a term. "Relevant" is a better description for what Google's PageRank delivers. As the company explains on its site:

> PageRank relies on the uniquely democratic nature of the web by using its vast link structure as an indicator of an individual page's value. In essence, Google interprets a link from page A to page B as a vote, by page A, for page B. But, Google looks at considerably more than the sheer volume of votes, or links a page receives; for example, it also analyzes the page that casts the vote. Votes cast by pages that are themselves "important" weigh more heavily and help to make other pages "important." Using these and other factors, Google provides its views on pages' relative importance.

Google doesn't view all links from all people equally. The more links you get to your site, the more your links to other sites are worth. Thus Google pays heed to those to whom we pay heed. Google realizes that trust is something we share with each other. Or put another way, any friend of ours is a friend of Google's.

Google found value in trust. Others are creating systems of trust as the core of their businesses. Facebook helps us build lists of those we know and trust. eBay turned internet commerce's disadvantage—fear of being robbed by merchants we do not know—into a unique opportunity by becoming the platform for trusted transactions of physical goods among strangers. Studies have shown that consumers are likely to pay higher prices to merchants they trust. Amazon, too, has created a system of trust in its reviews (though they can be infiltrated by both authors and their enemies) and in the money-where-your-mouth-is value of telling us that people who bought this also bought that. Prosper.com (which I'll discuss in the chapter, "The First Bank of Google") created a system of trust for person-to-person loans. PayPal did the same for person-to-person payments. We are witnessing the growth of the trust industry.

Social news service Digg has built a content community around trust. Users find and submit stories to the site, and then the community votes on what should go to the front page. That's editing by the mob and it works (especially if your interests gravitate toward the geeky). Instead of a staff, Digg has thousands of volunteer editors out there finding the interesting and noteworthy news on the web, and competing with each other to get it on Digg first. That makes the service lightning fast, a great source of alerts and updates. Diggers develop a reputation—anointed by fellow Diggers—by finding the most interesting stories fastest.

Journalists I know are suspicious of Digg and of the mob usurping their prerogatives and jobs. One day I sat at a lunch with a news executive and my son, Jake. The executive's a nice guy but not terribly interesting to a teen. So Jake had his nose buried in his iPhone as the executive belittled Digg, which he decreed to be over already. "Why would anyone trust this thing?" he asked. I turned to Jake and asked him what he was doing. "Oh, Digg," he said. As we quizzed him, Jake told the executive that he never goes directly to a brand like this man's newspaper or even to blogs he likes. He rarely types in one of those addresses and wonders what they have to tell him today. Mind you, he reads a lot of news—far more than I did at his age. But he goes to that news only via the links from Digg, friends' blogs, and Twitter. He travels all around an internet that is edited by his peers because he trusts them and knows they share his interests. The web of trust is built at eye-level, peer-to-peer.

Before I go on, let me acknowledge that, of course, things can go wrong. In 2005, the Los Angeles Times decided to be cyber-hip by inventing the "wikitorial," an editorial from the paper that the public was invited to rewrite. In no time, the quality of discourse around the first wikitorial descended to the level of that on a prison yard during a riot because the Times had made a fundamental error: A wiki is a tool used for collaboration, but there was no collaborating to be done on the topic of the Times' wikitorial—the Iraq war. I saw things going to hell and blogged that the Times would have been wiser to have created two wikis—one pro and one con—structured like an Oxford debate. The challenge to the opposing crowds would have been: Give us your best shots and let readers judge. It so happened that Jimmy Wales, founder of Wikipedia, saw my post and agreed. He headed to the Times to propose "forking" the wikitorial into two, but by then it was too late. The Times put a stake through the heart of the wikitorial. Since then, when newspaper people talk about interactivity, somebody will point to the danger of the wikitorial. Never mind that the form was misused; wikis now have cooties.

Interactivity has its limitations. Some people are simply wrong. Others are asses. Some need their meds. But don't let them ruin the party. Too often, I hear traditionalists in every industry suggest we throw out the internet baby with the bathwater: When they see one nasty comment, one hoax, one rumor, one lie, they try to use that to discredit the entirety of a service or of the internet as a whole. That's just as silly as wanting to ban phones, cars, or kitchen knives because something bad could be done with them. Of course, people misuse the internet. They misuse everything else, why should the internet be different? Where there is a challenge, though, find the opportunity. LiveWorld, for example, has made a business out of monitoring and maintaining communities.

Too many companies have been built not on trusting people but on making rules and prohibitions, telling customers what they cannot do, and penalizing them for doing wrong. Google has built its empire on trusting us. Trust Google on this.

Listen

At Google, we are God and our data is the Bible. It's through the data generated by our activity that Google listens to what we want, prefer, and need. Google vice president Marissa Mayer has said that Google is constantly trying to anticipate and interpret our desires so it can predict what we are going to do—our intent. It does that by watching our every move. When her team wonders whether a page should be this color or that, they don't make the decision themselves, nor do they hold a focus group. They put both colors on the web in an A/B test that measures which yields better usage. "We'll be able to scientifically and mathematically prove which one users seem to be responding to better," Mayer told students at Stanford, demonstrating an engineer's faith in numbers.

If a Google employee is meeting with Larry and Sergey to talk about users' needs, Mayer advised, she'd better come with more than her own conclusions. She'd better come with the data. "Their immediate question is, 'How many people did you test?'" This reliance on data as the proxy for the will of the people is so ingrained in Google's culture that it even supersedes organizational politics. "We rely so much on the data and we do so much measurement that you don't have to worry that your idea will get picked because you're the favorite," Mayer said. "Data is apolitical."

Google has faith in data because it has faith in us. When you take to heart the moral of James Surowiecki's 2004 book, *The Wisdom of Crowds*, you must realize that your crowd—your users, customers, voters, students, audience, neighbors—is wise. The next questions should be: How do you capture and act on that wisdom? How do you listen? How do you enable them to share their wisdom with each other and with you? How do you help them make you smarter (and why should they bother)? Do you have the systems in place to hear? Do you have the culture in place to act on what you hear?

The first answer is to listen before you speak. Many times, companies have told me they're going to blog to start conversations. Hold on, I tell them. Read before you write. Use search tools to find the conversations already going on about you and then join them. Look at every bit of data you have about how your constituents behave to learn more about their desires—and figure out what new data you can collect. Find ways to ask your public directly or through testing. If you're lucky, like Google, you will have the means to test the actions of thousands or millions of users in a day. About.com has 700 sites with useful information on very nichey topics and millions of users searching for answers in millions of articles. When I worked with them, I sat in metrics meetings while executives stared at no end of usage statistics projected on the screen, tracking the behavior of any and every link on all pages. They rigorously tested different versions of pages whenever they wanted to make a change.

Not every business and institution has the blessing of Google's and

About.com's data. Sometimes, of course, it's better to listen to people one-on-one, as Starbucks and Dell are doing with their give-us-your-ideas platforms and as countless companies do when they read blogs and forums. These methods beat focus groups and surveys, which pick people at random who may have nothing to say. It's better to listen to the people who have a reason to talk with you. Procter & Gamble chairman and CEO A.G. Lafley said in Strategy+Business magazine that he wanted customers to be "valued not just for their money, but as a rich source of information and direction."

Sometimes, listening itself becomes your product. Flickr listens well. The photo service founded by Caterina Fake and Stewart Butterfield and now owned by Yahoo created an incredible infrastructure to take in more than a million photos a day and enable users to organize them around captions and tags—one-word descriptions—which also enable fellow users to find them (and each other). This is all made possible, as we discussed earlier, because of Flickr's decision to make photos public by default.

Flickr brings out not just the wisdom of the crowd but also the aesthetic of the crowd and displays that for all of us to see. Go to flickr.com/explore/interesting/7days/ and press the reload button a few times or click the link there that enables you to view these images as a slideshow. I predict you won't be able to stop. It's mesmerizing. These are the photos Flickr has determined are interesting. How did they do that? By ranking popularity? No, that would likely lead to lots of pictures of thin young people who look good wearing very little on beaches—or, worse, to pictures of cute cats. Does Flickr do this with an army of editors? That would be the reflex of old media. But that would not scale, as they say in Silicon Valley; it would take a nation of editors to sift through the 3,000 pictures that come into Flickr every *minute*.

How does Flickr find interesting photos? Well, of course, they don't. We do. As Butterfield and Fake explained it to me, Flickr determines "interestingness" in a few ways. The first and most obvious component: Flickr measures the interactions—commenting, emailing, tagging, linking—that occur around a photo. Second, they map all these actions to see which users turn out to be hubs of activity. These people are presumed to be influencers and their actions are given extra weight because the Flickr community must trust them—a logic not unlike that used by Google's PageRank. Third, Flickr performs a reverse social analysis: If

Bob and Sally are emailing and commenting on each others' photos all the time, the system presumes they are relatives or friends; they have a social relationship built on familiarity. But if out of nowhere, Bob interacts with Jim's picture, the system then presumes that their relationship is based on the photo, not on life. The interestingness algorithm devalues Bob and Sally's social relationship and gives greater value to Bob and Jim's interaction around a photo. It's counterintuitive but sensible when you think about it.

Flickr ends up with a never-ending stream of interesting photos. Granted, being interesting is not as hard a test to pass as, say, relevance on Amazon or accuracy on Google. Still, look at Flickr's gallery. I'll bet you'll agree that almost all the choices are, indeed, interesting. Flickr is algorithmically aggregating the aesthetic of the crowd. Out of that comes a better service for every user, more opportunities to build traffic and revenue, a rich relationship of trust among those users and Flickr, and even new products. All from just listening.

New Ethic

Make mistakes well
Life is a beta
Be honest
Be transparent
Collaborate
Don't be evil

Make mistakes well

We are ashamed to make mistakes—as well we should be, yes? It's our job to get things right, right? So when we make mistakes, our instinct is to shrink into a ball and wish them away. Correcting errors, though necessary, is embarrassing.

But the truth about truth is itself counterintuitive: Corrections do not diminish credibility. Corrections enhance credibility. Standing up and admitting your errors makes you more believable; it gives your audience faith that you will right your future wrongs. When companies apologize for bad performance—as JetBlue did after keeping passengers on tarmacs for hours—that tells us that they know their performance wasn't up to their standard, and we have a better idea of the standard we should expect.

Being willing to be wrong is a key to innovation. Procter & Gamble's A.G. Lafley said in Strategy+Business that he improved the company's commercial success rate for new product launches from 15–20 percent to 50–60 percent, but he didn't want to push the rate higher because "we'll be tempted to err on the side of caution, playing it safe by focusing on innovations with little game-changing potential." Mistakes can be valuable; perfection is costly. The worst mistake is to act as if you don't make mistakes. That puts you on a pedestal, and when you fall off you better watch out: That first step's a doozie.

Consider Dan Rather. Minutes after he reported questions about George W. Bush's military service on CBS' *60 Minutes* in 2004, bloggers suspected that documents he had used as the basis of his story were faked. At the blog LittleGreenFootballs, Charles Johnson proved it. He took a memo supplied to Rather that reputedly had been typed using a 1970s-era IBM Selectric and then precisely re-created it using Microsoft Word on a next-century computer. He even made a neat animation that placed his document over the alleged original to show just how exact the match was. After his conclusions appeared on his blog, word flashed around the web. For 11 days, Rather ignored the ensuing storm, saying nothing. When he did respond, he dismissed his critics as political operatives. The smarter reply—the journalistically and intellectually honest approach—would have been to say, "Thanks guys. Let's share what we know and get to the truth together."

Rather came from an era of control when journalists were taught, ironically, to hide things from the public—sources, research, decision-making, and opinions. "Judge us by our product, not our process," a former network news president told me in a discussion about journalistic transparency at the Aspen Institute. But today, on the internet, the process has become the product. By revealing their work as it progresses, journalists can be transparent about how they operate and can open up the story for input from the public. Bloggers purposely post incomplete knowledge so they can get help to make it complete. Gawker Media publisher Nick Denton explains that such "half-baked posts" tell readers: "This is what we know. This is what we don't know. What do you know?" Corrections welcomed here.

I hear people fret that there are falsehoods and lies on the internet. There certainly are. And there are people who believe or want to believe those lies and errors. But there are also people such as Rather's bête-noire bloggers who are willing and able to ferret out facts. "We can fact-check your ass," blogger Ken Layne said in 2001. A lot of attention is given to the mistakes or sabotage we see on Wikipedia, but what's more impressive is watching the process of correcting and improving entries there, undertaken by people who get nothing out of it but the satisfaction of making things right. Snopes.com exists just to debunk urban legends. Wikileaks.org exists to give whistleblowers a place to share documentation of evildoings—and when a federal judge tried to shut it down in 2007, its community responded by replicating the site all over the web. Truth will out.

Contrast the Rather affair with the case of Reuters after one of its photographers was accused of doctoring a photo of Beirut during Israeli bombardment of the city in 2006. Some of the same bloggers, including Johnson, demonstrated that the photographer had used Photoshop to extend and darken a cloud of black smoke by copying and replicating parts of the picture. The wire service immediately pulled the photos and investigated the photographer's other work. Reuters fired him and changed its procedures to catch future tampering. Most important, Reuters thanked the bloggers, acknowledging that they, too, cared about the facts. *That* is how to make a mistake.

Life is a beta

Almost every new service Google issues is a beta—a test, an experiment, a work in progress, a half-baked product. It is a Silicon Valley punch line that Google products stay in beta forever—Google News was supposedly unfinished and in testing for more than three years—whereas Microsoft releases products and releases them again and releases them a third time before finally getting them (almost) right.

"Beta" is Google's way of never having to say they're sorry. It is also Google's way of saying, "There are sure to be mistakes here and so please help us find and fix them and improve the product. Tell us what you want it to be. Thanks." Most established companies would consider releasing unfinished products to market criminal: You can't produce a product that's not perfect—and not even done—or it will hurt the brand, right? Not if you make mistakes well. "Innovation, not instant perfect perfection," was Google vice president Marissa Mayer's advice to Stanford students. "The key is iteration. When you launch something, can you learn enough about the mistakes that you make and learn enough from your users that you ultimately iterate really quickly?" The internet makes iteration and development-on-the-fly possible.

Mayer put Google's worldview into cultural context: "I call this my Macs and Madonna theory. When you look at Apple and Madonna, they were cool in 1983, they're still cool today in 2006, 23 years later." How do they manage that? "They don't do it by being perfect every time. There's lots of missteps along the way. Apple had the Newton, Madonna has the sex book." When you make a mistake, Mayer advises, "you just iterate your way out of it or you reinvent yourself."

Mayer recounted a debate among engineers before the launch of Google News. Days before the start of the beta, they had enough time to implement one more feature—sort by date or by location—but couldn't decide. So they did neither. The day the service was released, they got 305 emails and 300 of them asked for sort by date. The users answered the engineers' question for them. "Just get the product out there and then have the users tell us where it is more important to spend our time." Google is not perfect. "We make mistakes every time, every day," Mayer confessed. "But if you launch things and iterate really quickly, people forget about those mistakes and have a lot of respect for how quickly you build the product up and make it better."

Google is unafraid of making mistakes that can cost money—courage one rarely sees in business. Advertising executive Sheryl Sandberg (who later was hired away from Google to be COO of Facebook) made an error she won't describe in detail that cost the company millions of dollars. "Bad decision, moved too quickly, no controls in place, wasted some money," she confessed to Fortune magazine. She apologized to boss Larry Page, who responded: "I'm so glad you made this mistake, because I want to run a company where we are moving too quickly and doing too much, not being too cautious and doing too little. If we don't have any of these mistakes, we're just not taking enough risk." Google CEO Eric Schmidt told The Economist that he urges employees: "Please fail very quickly—so that you can try again."

Facebook tends to blunder into new products, making mistakes as it goes. When Facebook introduced the news feed that compiles tidbits from friends' pages and activities, some users were freaked by what they perceived as a loss of privacy (even though anything going into news feeds was already public). Protest groups were formed inside the service, using Facebook to organize a fight against Facebook. Founder Mark Zuckerberg apologized for not warning users and explaining the feature well enough—communication was his real problem—and Facebook added new privacy controls. There was no exodus. Today, I don't think any user would disagree that the news feed is a brilliant insight; it is the heart of the service.

Though he makes mistakes, Zuckerberg makes them well by listening to customers and responding quickly. After a kerfuffle about a new Facebook advertising feature subsided, blogging venture capitalist Rick Segal begged us all to give Zuckerberg some slack. "He is going to make lots of mistakes, and he will continue to learn and grow. . . . We need to use care in beating up Zuckerberg and Facebook in general because we want these folks to push the limits of finding new ideas and trying to make sense out of all the data flowing everywhere. Try it and get some reactions, adjust, find the happy center, rinse and repeat. . . . If they do really bad things, people vote with the mouse clicks." It's not the mistake that matters but what you do about it.

Be honest

Fake news anchor Jon Stewart is one of the most trusted newsmen in America because he calls bullshit. Howard Stern is the king of all media in the U.S. because he's honest. The tagline of Stern's personal news service on satellite radio: "No more bullshit." Shouldn't that be every news organization's tagline? Every company's?

I've been a fan of Stern's since I reviewed his show for TV Guide in 1996 and discovered, counterintuitively, that he is best taken not in small doses but in large doses. If all you heard of him were the odd belch, you'd be forgiven for dismissing him. But Stern is greater than the sum of his farts. Listen for a few days and you will hear the rare man—rare especially on broadcast—who is not afraid to say what he thinks and what we think but don't dare say. In the plasticized, packaged world of roboreporters on TV and shtickmeisters on radio, it's a relief to hear somebody who's candid, honest, and blunt. He is open and transparent about his life. He is unafraid to ask the tough question; I only wish that the PR-laden morning shows were as direct as Stern or as skeptical as Stewart.

Stewart, anchor of Comedy Central's *The Daily Show*, came in fourth among the most admired newsmen in America, tied in that slot with network anchormen Brian Williams, Tom Brokaw, Dan Rather, and Anderson

Cooper in a 2007 survey conducted by the Pew Research Center for the People & the Press. Stewart's spin-off, Stephen Colbert's *Colbert Report*, mocks spin, shooting buckshot into the pomposity of news shows, talk shows, pundits, and PR.

Stern, Stewart, Colbert, and bloggers everywhere say what they think. In them, we hear the language of the internet age: honest, direct, blunt, to the point, no bullshit, few apologies. Their tone may shock old, controlled sensibilities. But complaining about it, tsk-tsking it, trying to clean it up, or trying to ignore it won't work. The post-media generation raised on honesty and directness expects truth and bluntness from others. With Google, it is harder to hide behind spin, to control information, or to hope that people will forget what you said yesterday or the mistakes you make today. The truth is a click away.

Institutions are learning to acknowledge their mistakes and apologize. When he took office following predecessor Eliot Spitzer's sex scandal, New York Governor David Paterson preemptively admitted having an affair, among other peccadilloes. Apple had a near-disaster in the launch of its Mobile.me service and Steve Jobs admitted it publicly. This is honest talk, which comes in a human voice. Even in the machine age—the Google age—that voice will emerge and succeed over a filtered, packaged, institutional tone. *The Cluetrain Manifesto* (which you can read for free at Cluetrain.org) teaches this lesson in its 95 theses, which begin:

1. Markets are conversations.

2. Markets consist of human beings, not demographic sectors.

3. Conversations among human beings *sound* human. They are conducted in a human voice.

4. Whether delivering information, opinions, perspectives, dissenting arguments, or humorous asides, the human voice is typically open, natural, uncontrived.

5. People recognize each other as such from the sound of this voice.

6. The internet is enabling conversations among human beings that were simply not possible in the era of mass media.

In every interaction you have with your constituents, speak with a human voice as if you were speaking face-to-face. Be boldly, bluntly honest when admitting your mistakes—and when disagreeing with the public. Lock your PR people away. And remember, everything you say is searchable. Think of Google as the angel on your shoulder keeping you honest.

Be transparent

My life is an open blog. On the "about" page on my site, I try to practice what I preach about transparency. I reveal my business relationships: the companies for which I work, write, speak, and consult. I reveal personal relationships: companies where I used to work, where I have friends, and even where I have been turned down for jobs. I list stocks I own. I sometimes write about religion, so I reveal mine. As I often write about politics, I reveal my views and—to the horror of traditional journalists—my votes. This page is my defense against an accusation that I might try to hide affiliations, opinions, or conflicts of interest. At the end of this book, I will also make relevant disclosures.

I'll throw out this challenge to you in your organization: Why keep secrets? Or why keep more secrets than you have to? I've heard the argument: Your competitors will steal good ideas. But transparency will build a relationship of trust with your constituents and open up new opportunities. The ethic of transparency sums up much of what has come before in this book: the need to involve your constituents in your process, the need to hand over control through openness and information, the benefits of open-source networks, the benefits of the gift economy, the ability to listen.

But I must acknowledge the irony of advocating transparency in a book about Google, which in many ways is as opaque and secretive as Dick Cheney. You can't get into a Google office without signing a nondisclosure agreement. Google won't reveal details of its revenue split with sites that run its ads. It refuses to list its Google News sources. It won't tell us how many servers it has. It chooses not to use open-source software for some functions, like managing its cloud of computers, so it can retain a proprietary advantage.

Still, as we've just discussed, Google does develop most of its products in public by releasing unfinished versions and getting help from users. In that sense, it is unusually transparent, willing to work in the open and involve its users in development. I suggest you follow Google's example in its product development and ignore its silence and opaqueness elsewhere.

Collaborate

If you don't open up, you can't collaborate. Collaboration with customers is the highest and most rewarding form of interactivity, for that is when the public tells you what they want in a product before you've made it. If you're lucky, they'll take ownership in the product you create together. They won't just buy it, they'll also brag about it.

I have tried to make this book collaborative. I didn't put chapters online as I turned them out to have readers correct and edit them, as other authors have done; that is too after-the-fact. Nor did I try to make the book a product of democracy ("vote on what I should say"); deciding what to say is, in the end, my job. Instead, I discussed ideas in the book on my blog as I researched them and thought them through and asked readers for guidance, which they generously gave. The chapter "Google Mutual Insurance" that follows is a product purely of that discussion.

Collaboration is good business. Michael Dell spoke to me about "co-creation of products and services," a radical notion from a big company whose policy had once been to look at and not touch its blogging customers. Now it tries to make, change, and support products collaboratively. "I'm sure there's a lot of things that I can't even imagine but our customers can imagine," Dell said. "A company this size is not going to be about a couple of people coming up with ideas. It's going to be about millions of people and harnessing the power of those ideas." Once you can hear them.

Start by letting your customers into the genesis of your products: your design process. Impossible, you protest: It's a secret. Well, why is that? By closing off design, you're also closing yourself off from the best ideas of the people who need, buy, and care about your product. Think how much more valuable your products and company would be if you were to give your customers exactly what they want. Take one project or product and try being radically transparent about it (as we will explore in the chapter, "Manufacturing"). Blog about your plans and decisions. Join in conversations—human conversations—with customers. Ask people what you should do. Admit mistakes. Open up.

Your competitive advantage is not that your designs are secret but that you have a strong relationship with your community of customers. I'm not suggesting that you hand over design to a committee of the whole. That would be like turning over the boardroom to a giant focus group. Design cannot come out of town-hall meetings. It's still your job to come up with good ideas, to invent, inspire, surprise—and to execute well. Companies are not democracies. But neither should they be dictatorships. They should be—but too rarely are—meritocracies. Your challenge is to get good ideas to surface and survive from within and without and to enable customers and employees to improve your ideas and products.

Don't be evil

We can't leave a chapter about ethics and Google without addressing its famous self-admonition: "Don't be evil." Larry Page and Sergey Brin interpreted the pledge this way in a letter they wrote before their 2004 initial public offering: "We believe strongly that in the long term, we will be better served—as shareholders and in all other ways—by a company that does good things for the world even if we forego some short-term gains. This is an important aspect of our culture and is broadly shared within the company."

They defined good behavior as delivering unbiased search results and not accepting payment for advantage in listings. They vowed to clearly label advertising, comparing their policy with newspapers' rules. They set themselves apart from marketers, saying: "We believe it is important for everyone to have access to the best information and research, not only to the information people pay for you to see."

One could see their covenant on evil either as the height of hubris—Google declaring itself the headquarters of corporate virtue—or as a case of saying what should be assumed. It necessarily raises questions about whether Google is living up to its credo. Google has censored search results in China, arguing that it is better to bring a hampered internet there than no internet at all. I don't agree and believe that Google has more power than it knows to pressure countries around the world to respect openness and free speech. Google, like Yahoo, has handed over information to governments—Google in India, Yahoo in China—that led to users being arrested simply for what they said. As an American and a First

Amendment absolutist, I'd call that evil. I think that Google's lack of transparency about advertising splits is not evil but is also not virtuous business. Some would argue that Google is the bad guy for making money off news headlines while news organizations are struggling; I disagree and say that Google is doing news sites the favor of sharing audience. Some would say that Google can do evil with the private information it has about our searches, clicks, and even health history; I don't think we've seen evidence of misuse yet.

Is Google a monopoly? In 2008, as the U.S. Justice Department began an antitrust inquiry into Google's deal to sell ads for Yahoo, New York Times columnist Joe Nocera reported that Sourcetool.com had filed a complaint against Google for raising the company's ad rates prohibitively high. Google's algorithms and employees found that Sourcetool did not meet its standards; it resembled a spam site, whether or not it was one. The rate increase was Google's way of shooing off the site. Sourcetool disagreed and said Google was ruining its legitimate business. The implication was that Google could wield the power of the monopoly. But in the Google age, nothing is as it seems. The issue is not that Google is a monopoly but that it has become *the* marketplace—the best place for us to find information and for advertisers to find us—as newspapers were in their time and as craigslist is today. Marketplaces have the power to unilaterally charge what the market will bear. craigslist sets most of its ad rates to zero. Google says it doesn't set rates but enables the market to do the job in auctions. Except in the case of Sourcetool, Google did unilaterally set the rate. The question is whether we trust Google with the power to do that. Is Google a monopoly? Not yet.

The next question is whether Google can live by its golden rule as it grows huge and gangly—as middle managers start second-guessing their bosses, as bonuses and greed or simple self-interest overtake the gospel according to Google. Time will tell.

Is Google evil then? On balance, I don't think so. But its day is still young. At least Google is trying to be good. That's more than one can say for some companies I'm sure we both could name. Wouldn't other companies do well to make the same pledge on evil? It should be chiseled over doors on Wall Street. If only, in the poisoned process that led to the financial crisis of 2008, enough people had asked whether seeking and issuing toxic mortgages and making and selling toxic assets were evil—instead of someone else's problem—I wonder whether we'd have reached that nadir.

Imagine if in cable company meetings on pricing and bundling or restricting internet access someone were to ask: Is this the best we can do for our customers? Are we exploiting them? Is this evil? Imagine if someone were to ask at the meeting where airlines chose to fight a New York State law requiring that passengers be given clean air and water: Is this any way to treat our passengers? Aren't we being evil? I wouldn't much like to be that person—Mr. Goody-Goody, director of whistle-blowing, vice president of virtue. But I do believe that if companies were to ask themselves—and employees were empowered to ask—whether they were being good or evil to their customers and communities, they would often make different decisions. It's not a bad rule.

Wal-Mart made news early in 2008 when it sued a former employee who had been hit by a truck and left severely brain-damaged. The store wanted to recoup what it had paid for her care after she won a $1 million judgment against the trucking company. After legal fees, the victim received $417,000; Wal-Mart sued to recover $470,000, which would have left the employee's family with nothing to pay for nursing-home care. Wal-Mart was apparently within its legal and contractual rights to recover money; that's what the fine print said. But if just one person had asked the right question in the memos or meetings about this case—Is this evil?—the company would have saved itself horrible publicity on network news, in papers, and in blogs from people who used the story as exhibit A, proof that Wal-Mart is evil. Eventually, the company backed down and did the right thing: It dropped its suit against the brain-damaged woman. But the PR damage cost more than the money at stake. "Don't be evil" is good business.

That was the point made by Umair Haque as he excoriated Facebook later in 2008 for preventing Google from using Facebook members' data (with their consent). On his Harvard Business Review blog, Haque called Facebook evil. That's a bit strong, I'd say, but he was making a business point: "What's really going on here? There's a massive tectonic shift rocking the economic landscape. All these players are discovering that the boardroom's first and most important task is simply to try always and everywhere [to] do less evil. In the dismal language of economics: as interaction explodes, the costs of evil are starting to outweigh the benefits."

Let's repeat that and dub it Haque's Law: As interaction explodes, the costs of evil are starting to outweigh the benefits. That, I think, is what Google is talking about when it promises not to be evil. It is not a campaign pledge or a geeky Bible lesson about good and bad. It is a calculated business rule: When people can openly talk with, about, and around you, screwing them is no longer a valid business strategy.

New Speed

Answers are instantaneous
Life is live
Mobs form in a flash

Answers are instantaneous

Google has spoiled us rotten. Think back to the time before Google—it was only a decade ago—and remember the mines you had to dig to find any bit of information. Good God, we actually went to libraries. We waited for answers and went without them. Now I ask Google a question, any question, and it brags that it has given me the answer in fractions of a second. I wanted to tell you just how fast that is compared to, say, the blink of an eye. So what did I do? Of course, I asked Google how fast an eye blinks and in .3 seconds it told me that a blink takes .3 seconds.

One of Google's own principles—the "10 things Google has found to be true"—is: "Fast is better than slow." A pillar of its design principles—from Google's list of what makes a design Googley—is: "Every millisecond counts. . . . Speed is a boon to users. It is also a competitive advantage that Google doesn't sacrifice without good reason." Speed is a tenet of the Google religion.

Google has made us an impatient people, more than we know. If we can get any of the world's knowledge in a blink, why should we wait on hold or in line or until your office opens? Why should anyone give us incomplete information when completeness is a search away? We want what we want, and there's no reason we shouldn't have it—now.

Every industry is affected by this new speed. Fashion—as practiced in international chains such as Zara and H&M—reacts to new styles overnight. A trend comes off the runway and it's imitated—flattered, that is—in a flash. Information on what is and isn't selling is fed back constantly so stores can adjust their stock and even the companies' manufacturing and design. Speed becomes not only a competitive advantage but also a strategic necessity. The more quickly businesses can adjust to customers' actions and desires—the more quickly they can learn from them and try to stay ahead of them—the better business will be.

A lack of speed is a strategic disadvantage. Many industries are saddled with slowness because they are trapped by atoms and complexity. Automobiles are fashion products but because their machinery and supply chains are so complex, they cannot exploit new trends (and gas prices) until the trends are already out of date. (Is there an alternative? I'll brainstorm about that in the chapter, "The Googlemobile.")

The book publishing industry is shamefully slow. I negotiated the contract for this book about a year before you got it in your hands (and by the way, I've been meaning to thank you for picking it up). That's damned speedy for a book. As other forms of knowledge, entertainment, and content creation speed up, so must books. (I'll explore that, too, in the chapter, "GoogleCollins.")

Education prides itself on not being speedy. As an academic, I appreciate the virtues of deliberation, of ideas being reviewed and challenged, of knowledge fermenting over time. But those of us who teach students in rapidly changing arenas (I teach digital journalism) must get better at keeping up with—no, at getting ahead of—our students, industry, and society.

Perhaps only religion can claim exemption from the imperative for speed. If any institution relies more on permanence than hastiness, God's does.

Google, like God, values permanence. In its search results, Google gives more credence to sites that have been online long enough to build a reputation over time via clicks and links—this is the essence of Page-Rank. As a result, Google's search has been better at delivering completeness and relevance than currency. Google is not great at surfacing the latest links on a topic. Google has fresh links in its database because it constantly and quickly scrapes the web to find the latest content, but until those new entrants gather more links and clicks, it's hard for Google's algorithms to know what to make of them. Could this be a chink in Google's armor?

Life is live

Just as Google and the rest of us start to get our hands around currency—finding the latest—the web speeds up even more. The internet is going live.

I have broadcast live video to the world on the internet from my Nokia phone—no satellite truck, no microwave hookups, no broadcast tower or cable company, just me and my phone, live. The next time a big news event happens—a 9/11 in New York, a 7/7 in London, or an earthquake in China—witnesses will have the ability not only to capture but also to share with the world what they see as they see it.

Live video from witnesses will have a profound impact on news networks. They have begged witnesses to send in their tips, photos, and videos—after the fact. When a student at Virginia Tech University went on a shooting spree in 2007, a fellow student recorded the sound of the shots with his camera-phone. He sent the video to CNN, which took more than an hour to vet it and get it on-air and online. If that student had been broadcasting using a phone on live video services such as Qik .com and Flixwagon.com, he wouldn't have sent anything to CNN but would have been sharing the video on his own. CNN's choice would be whether to link to the student's broadcast or embed it on its web page or in its broadcast. It could not delay the decision, for then the live video would not have been live anymore.

When China's Sichuan Province suffered its horrendous earthquake in May 2008, people who felt it firsthand shared their experience via Twitter, a microblogging platform that enables users to send and receive 140-character-long updates to friends who follow them on the web or via short-message services on their mobile phones. Twitter was cofounded by Evan Williams, one of the creators of the company that built Blogger, which revolutionized publishing. Now he has taken publishing mobile and live. I was shocked that this service, just two years old, had spread to China—but then, I, too, sometimes forget the internet's ability to spread in an instant, distance be damned. What isn't shocking is that people in the quake zone would use Twitter to update friends. That's what it is made for. If I were going through a quake, I'd want to tell family and friends that I was safe, wouldn't you?

Twitter is becoming the canary in the news coal mine. Developers at the BBC and Reuters picked up on Twitter's potential and created applications to monitor it for news catchwords such as "earthquake" and "evacuate." Journalists search Twitter to find witnesses to interview and quote. During the Sichuan quakes, Twitter user casperodj wrote, "CREEPY! while i'm typing, there's an aftershock hitting!" News organizations also search Flickr, YouTube, Facebook, and blogs to find photos and videos that witnesses record, long before professional photographers arrive.

Imagine the problem the live web presents to Google: How can it search for and find things as they are happening? Oddly, Wikipedia can be quicker at updating current information than otherwise-speedy Google. It carried the news of Tim Russert's and Paul Newman's deaths before major news sites. During momentous events such as the 2004 tsunami, Wikipedians maintained entries with up-to-the-minute news. In the John Henry duel of man and machine, it's nice to see man winning one. Perhaps we need more human-powered means of recognizing what's new and what's hot—that is what the search service Mahalo contends and that is a core value of human-powered aggregator Digg. There is a business opportunity in finding currency—complementing Google's completeness—for news organizations, industry trade groups, aggregators, and bloggers.

Live brings an important benefit to the web: It makes the internet interactive, person-to-person, nose-to-nose. When something is happening live online, we can have conversations around it, we can share the same experience and discuss it, we can influence events. But it also makes the web perilous for businesses being talked about—unless they have the facility to listen to and join the conversation as it happens.

Mobs form in a flash

In this live connection machine, people of similar interests and goals—call them communities or call them mobs—can find each other, coalesce, organize, and act in an instant. Howard Rheingold dubbed them *Smart Mobs* in the title of his 2002 book. Rheingold chronicled the fall of Philippine president Joseph Estrada at the hands of a smart mob of tens of thousands who were gathered together in only an hour by SMS messages on phones that told them to "Go 2 EDSA," an address in Manila, and to "Wear blck."

On a much less grand and profound scale, I watched Twitter form mobs at the South by Southwest conference in Austin in 2008 after attendees excitedly swarmed to the most anticipated party—Google's, of course—only to find a line three geeks thick running three blocks long. One of those would-be partiers, Gary Vaynerchuk, a tech-savvy wine merchant and video blogger you'll hear from later, in the chapter on retail, decided to chuck the Google party and make his own. He used his phone to send a message to Twitter asking who wanted to join him. Vaynerchuk already had a few thousand friends following him and scores of them were in Austin. It helped that Vaynerchuk had shipped a few cases of good wine to Texas. A party formed. On Twitter, I watched as one and then another and then another of his friends told their friends they were heading to the party. It came together in minutes.

Not long after this episode, I saw tech blogger Michael Arrington, who runs the powerful TechCrunch.com, complaining loudly on Twitter—as best he could in 140-character bursts—that his Comcast internet connection had been down for 36 hours. He gave us a serial narrative about his time on hold and how he was told this was a California-wide issue (though fellow Californians replied on Twitter that they had no problems). Arrington went to a friend's house to get on the internet and Twittered that he would use his blog to make Comcast miserable. I linked to this on my blog and speculated that with Arrington's reach, he'd gather a Twitter mob in an instant. Something surprising happened instead: Comcast called Arrington and sent technicians out to fix the problem. They had monitored Twitter and read about his difficulty. Other bloggers and Twitterers were dubious and said so, but a Comcast rep responded to them on Twitter, proving he was there and listening. Comcast knows that it has to be on top of the conversation as it happens. Every second counts.

The internet has caused you to lose control of so much—brand, message, price, competition, secrecy—but more than anything, you've lost control of timing. You can no longer decide when to put your story out or when to answer critics. You can't subject your customers to waiting on hold—no matter how often you tell them that their call matters to you—without them complaining, revolting, and leaving quickly and publicly. The idea of holding back products and popping them out as surprises insults your customers (well, unless you're Apple). The earlier they're involved in your process, the better. The internet has changed the speed, the rhythm, and the process of business and next will do the same to government.

When customers come looking for you on Google, you'd better have answers to their questions on your web site before they are asked. When customers talk about you in public, you'd better have the means to hear and respond. It's simple for a competitor with a better answer to steal your customers in a flash.

New Imperatives

Beware the cash cow in the coal mine
Encourage, enable, and protect innovation
Simplify, simplify
Get out of the way

Beware the cash cow in the coal mine

Sometimes, success can blind you to the oncoming possibility of failure. And fear of failure can keep you from success.

When I was TV critic for TV Guide in the mid-1990s, it still sold more copies in a year than any magazine in America. But it was slowly fading, stuck in the first stage of death: denial. Its circulation had fallen from more than 17 million a week to 15 million, then 13 million while I was there (entirely the fault of my bad taste, of course). TV Guide couldn't keep pace with the explosion of television: Dozens, then hundreds of channels wouldn't fit on the magazine's little pages. The editors tried more than once to produce a larger version with big, colorful grids, but the old readership of the magazine was stuck in its ways, addicted to listings. There was the other problem: The readers were old and getting older. As I remember, when one readership survey came back with less than the usual level of response, a follow-up study was performed to find out why people hadn't completed their questionnaires. The answer: Most of the folks who hadn't responded had died.

Meanwhile, competition only grew. Listings were appearing on TV and computer screens, forcing TV Guide to enter those businesses. Newspapers' TV listings had long been perceived as free by readers. There was discussion of syndicating TV Guide's listings to papers—which, using Googlethink, could have spread the brand—but the magazine feared that would cannibalize the core product. Beware any strategy built on protection from cannibalization, for it probably means the cannibals are at the door and ready to eat you for lunch.

Fast-forward a dozen years, long after I'd left. In 2005, TV Guide transformed into a regular-sized magazine with big, colorful grids. At the same time, it eliminated almost all of its 140 local editions. It raised its price. And it lowered the circulation it guaranteed advertisers to 3.2 million, a dizzying drop from its high of 17 million. About then, I had lunch with my old boss from TV Guide, who had also moved on. I said the company had finally done everything we should have done a dozen years before: putting out the right product, reducing costs, and getting realistic about its legitimate circulation. Why didn't we do that? I asked rhetorically. She responded: "You know why. Because it was a cash cow."

Cash flow can blind you to the strategic necessity of change, tough decisions, and innovation. Take the fate of TV Guide as a warning: Beware the cash cow in the coal mine.

How many companies and industries fail to heed the warnings they know are there but refuse to see? The music industry is, of course, the best example of digital dead meat. Detroit waited far too long to make smaller cars and pursue electricity as a fuel. Many retail chains opened stores online but stopped there, not seeing opportunities to forge new relationships with customers as Amazon had. Telecom companies were blindsided by the emergence of open networks that undercut their businesses—even though those networks operated on the telecom companies' own wires. Ad agencies kept trying to forestall the reinvention of their industry, still buying mass media even as more targeted and efficient opportunities grew on the internet. News executives thought they could avoid change and even believed they should be immune from it because they were holders of a holy flame: Journalism with a capital J. They finally woke up when they watched the giant Knight Ridder chain get gobbled up by the McClatchy chain, which like every public company in the sector lost billions in market value. Now newsmen are willing to change, but it may be too late for them—as it was for the one-time giant TV Guide. They lost the next generations of customers. They lost their destinies because they wanted to save their pasts. Protection is not a strategy for the future.

Encourage, enable, and protect innovation

Google is well-known for giving its technical employees the chance to use 20 percent of their time to work on new ideas, new products, and new businesses. "A license to pursue your dreams," is what Google's Marissa Mayer called the policy in Fast Company magazine. A 2008 article in the Harvard Business Review by Bala Iyer and Thomas H. Davenport quoted a Google employee explaining on his blog: "This isn't a matter of doing something in your spare time, but more of actively making time for it. Heck, I don't have a good 20 percent project yet and I need one. If I don't come up with something I'm sure it could negatively impact my review." Google *requires* employees to innovate. It's part of the job. It's how workers are valued. It's how Google grows. In 2006, Mayer said that half the new products and features launched by Google in the last six months of 2005 came from work done under the 20 percent rule.

I'm not saying that every company is like Google and could or should implement its 20 percent rule. Even Google doesn't extend the offer and expectation to all its employees (Iyer and Davenport say that's a mistake). I understand how this policy could be impractical. Maybe you've already cut so close to the bone that you fear this reallocation of time and productivity could throw you over the edge. Maybe your employees aren't built to invent—after all, not every company is populated with Ph.D.s in rocket science like Google is.

But anyone anywhere in a company could have a brilliant idea. How do you hear it? How do your employees propose new products, methods, or systems—through the dead-end suggestion box? How will they be rewarded for innovating? Who will try to stop them? Do you have a culture of innovation or is this just something you say at management meetings?

You need to encourage employees to suggest new ideas—even suggestions that will cannibalize, destroy, and rethink your business. It's better for you to disrupt and cannibalize yourself than for a competitor to do it to you. Just as Dell, Starbucks, and Salesforce.com maintain versions of their ideas platforms for employees and as Best Buy has BlueShirt Nation, its online community where employees solve problems, so does Google maintain a place for ideas. "It's like a voting pool where you can say how good or bad you think an idea is," Mayer told Fast Company. "Those comments lead to new ideas." Add the lessons of openness and transparency to the need for innovation and you will end up building spaces where employees can share ideas and improve them. Procter & Gamble's A.G. Lafley said in Strategy + Business that "a practice of *open* innovation" (his emphasis) with "a broad network of social interactions" is critical. "The idea for a new product may spring from the mind of an individual, but only a collective effort can carry that idea through prototyping and launch."

There are different schools of thought about ownership of ideas. Lafley emphasized collective effort. Mayer said that in Google's "incredibly open culture" the company tries to avoid territoriality and to "give ideas credit, not credit for ideas." Nike understands the need to own and protect an idea. In 2008, I attended a brainstorming session at Nike aimed at finding better systems to encourage employees to contribute to their communities. We heard from four employees who were already doing this on their own. One of them, footwear designer D'Wayne Edwards, created a contest for young and aspiring designers. Edwards had many agendas: He wanted the company to nurture designers like him. He had grown up dreaming of creating sports footwear. Mentors gave him guidance and breaks, which led to his dream job at Nike. He wanted to give back. He also believed that Nike owed a debt to the urban young people who had elevated the company's brands and made them hot. And he believed his contest would help the company find and develop talent with new ideas. Edwards said the contest's winners, though just teenagers, had enough talent and innovative spirit to start working at Nike the next day.

The group discussed Edwards' process in hopes of replicating—or at least not ruining—his innovation and enthusiasm. We heard his fears. Edwards didn't want to be stopped, so he didn't start by asking permission. At Nike, they said, employees are told it's better to seek forgiveness than permission. Just do it, you know. Edwards also didn't want someone taking over his project, taking credit for it or corrupting it; he demanded ownership. Edwards needed Nike because its brand would attract young people and inspire them. Nike was his platform. To use the brand, he had to get his project vetted by lawyers, but he picked ones he knew would help. It was a rogue operation—innovation is, by definition, rogue. So here was Nike convening a meeting of insiders and outsiders to figure out how to nurture more rogues.

Bureaucracies, task forces, org charts, and formal processes do not breed innovation. They kill it. When I came up with the idea for Entertainment Weekly magazine at Time Inc. in 1984, it was rejected out of hand because the company's top editor did not think one magazine could possibly serve people who liked movies, TV, books, music, and video. People who watch TV, he said, do not read books. Six years later, my proposal arose again from the dead-idea file. After I made prototypes aplenty and we tested the idea exhaustively and made business plans galore, a task force was assembled for the express purpose of trying to kill it. You could say they were there to perform due diligence or you could say they were there to cover the bosses' asses. In any case, the magazine finally started. After an astounding $200 million investment—not all of it my fault—Entertainment Weekly became a franchise that brings in a few hundred million dollars a year. Innovation happens in spite of the structure of organizations.

In 2008, I joined a seminar on innovation at the World Economic Forum at Davos. It was a highly formatted hour, with the entire room sitting in a circle (making the moderator dizzy). They had us write down the technology we loved most. Then we compared notes with a neighbor and came up with some neat invention out of this mashup. We heard a few cute ideas and then, thank goodness, a scientist in the room put a stop to it. This, he said, is not how innovation is made. Scientists start with a problem and then try to find a solution. I'll show in a later chapter on the Google.org foundation, "Google Power & Light," that Google's founders approach invention in that order: first find the problem, and then create the solution. Beware the cool idea.

Of course, innovation and ideas do not come only from within. Remember Michael Dell saying that a company cannot be built on the ideas of a few people. "Ideas come from everywhere," Mayer told Stanford's students. When Google got into mapping, she said, it found engineers in Australia "who were just amazingly good at mapping interfaces" and then hired them. Google bought other products and ideas this way, leading to platforms for blogging, feeds, Google Docs, and advertising systems. Don Tapscott, author of *Wikinomics*, told the BBC's Peter Day on his *In Business* program in 2007 that Procter & Gamble now relies on ideas and solutions not invented there but "proudly found elsewhere."

Day went on to report on the solutions platform InnoCentive, where scores of companies post problems with offers of rewards for solutions from independent inventors, scientists, and tinkerers, whom InnoCentive calls "solvers." The problems range from the profound (a $1 million reward to find "a biomarker for measuring disease progression in Amyotrophic Lateral Sclerosis [ALS or Lou Gehrig's Disease]") to the scientifically geeky ("near complete conversion of phenol compounds into non-volatile or in-soluble products in an aqueous solution") to the prosaic (a large company wanted "bakeable cheese technology" for snack products; another offered $5,000 for "novel approaches to gently and effectively clean a baby;" and the Rockefeller Foundation offered $20,000 for the design of solar-powered internet routers).

No matter where the ideas come from, innovation is, of course, all about people, their talent, and how you nurture it. Rishad Tobaccowala, an advertising executive you'll hear from in the chapter, "Advertising," says the genius of Google's 20 percent rule is that if you enable people to follow their passions, they'll as much as work for free. Google, he said, knows how to find smart people and give them what they want: "They want to work for a winner. They want to have the ability to feel special. They want to have the ability to follow their passions. They want the ability to make money. And they want to have the ability to increase their external market value—it helps me build a brand: myself." Mayer said just working with smart people challenges other people in the company—and besides, they're easier to manage.

Google's lesson is clear: Make innovation your business.

Simplify, simplify

Once you decide what business you're really in, once you settle on your strategy, once you figure out how to execute it in the new architecture and realities of the Google age, once you cast a new relationship with your world, once you absorb new ethics of this new era into your company's culture, once you make innovation a keystone of that culture, then there's one more important thing to do, another vital lesson to learn from Google: simplify.

In their 2005 history of Google, authors David A. Vise and Mark Mal-seed recounted the story of Google testing an early version of its spare and spartan home page with users:

The testers were told to use Google to find the answer to a trivia question: Which country won the most gold medals in the 1994 Olympics? They typed www.google.com, watched the homepage come up on the screen, and then they waited. Fifteen seconds went by . . . twenty seconds . . . forty-five. [Marissa] Mayer wondered what was going on, but didn't want to interfere. Finally, she asked them, "What are you waiting for?" "The rest of the page to load," they answered. The same thing kept happening all day, Mayer recalled. "The web was so full of things that moved and flashed and blinked and made you punch the monkey that they were waiting for the rest of it to show up."

Mayer's team changed the design to make the copyright notice at the bottom of the page stand out, just to let users know the page was done loading so they could get started.

I had to learn the lesson of simplicity myself when I debated about the title of this book with my editor and publisher. My original title was *WWGD? (What Would Google Do?)*. It was a joke that I knew would work only in America, inspired by bumper stickers and bracelets that ask, WWJD? (What would Jesus do?) In this equation Google was God. But the publishing company thought this double title was too complicated. They wanted to simplify. I argued with them, holding dearly to my gag. To my editor's discomfort, I decided to take the debate to my blog readers, as is my reflex now. A great discussion ensued with a few dozen comments, a majority disagreeing with my argument. Then a commenter named Ellen advised: "To me, it doesn't matter what we all think. You should decide based on What Google Would Do, since that is the point of your book." Right. What would Google do? It would simplify. I had to follow my own rules. I simplified the title.

Google is perhaps the most powerful single tool that can be used by anyone on earth. But it is also the simplest. Compare Google's home page with a TV remote control, a clock radio, a tax form, an insurance policy, any legal document, many ecommerce sites, Microsoft Word's toolbar, most companies' org charts, and the last five memos you wrote. Google is simple.

Google shares its design aspirations on the web for all to see. "The Google User Experience team aims to create designs that are useful, fast, simple, engaging, innovative, universal, profitable, beautiful, trustworthy, and personable," the team wrote on a Google blog. "Achieving a harmonious balance of these 10 principles is a constant challenge. A product that gets the balance right is 'Googley'—and will satisfy and delight people all over the world." Their key design principle: "Simplicity is powerful. . . . Google teams think twice before sacrificing simplicity in pursuit of a less important feature." Or to quote another internal Google company principle: "It's best to do one thing really, really well."

Simplicity is empowering. I don't have to use Google the way someone else says I should, following its path of navigation. I don't have to feel stupid looking up instructions. Google never makes me feel foolish for making mistakes ("Did you mean . . . ?" it graciously asks when I misspell or mistype). It doesn't waste my time trying to find what I want. It just gives me a blank box and puts the world behind it.

Design is about more than aesthetics. Design is an ethic. Design is the path by which you interact with your public. Magazines, clothes, and cars aren't the only things that are designed. Companies are designed. Services are designed. Rules are designed. The simpler and clearer the design, the better. To be simple is to be direct. To be direct is to be honest. To be honest is to be human. To be human is to be in a conversation. To be in a conversation is to collaborate. To collaborate is to hand over control. And we are back to where we started, to Jarvis' First Law: Give the people control and we will use it. Don't and you will lose us. Simple.

Get out of the way

One more law from one more leader.

Craig Newmark, founder of craigslist, is a wonderful character. You'll never meet a more unassuming revolutionary and mogul. Proudly geeky, rarely fashionable (the one time I saw him in a tie, he said he put it on just to scare people), soft-spoken, impish, and ironic, he is not what you expect, whatever that is.

Newmark confounds business people, running a service mostly for free. He charges just for job ads and real-estate ads in some cities. By various accounts, he has destroyed a few billion dollars in value in the newspaper classified business. But as I said earlier, Newmark isn't at fault—the internet is. craigslist reveals no figures publicly, but it has been estimated that it brings in $100 million a year with just 25 employees. Still, money does not motivate Newmark and his president, Jim Buckmaster. Newmark could exploit his service for many times more income and equity. He could sell out for a fortune. He has no intention of doing either. That's what business people don't understand about him. He is like an alien to them (indeed, he is a bit like ALF).

Newmark introduces himself as the founder and customer service rep for craigslist. That always gets a laugh, but he doesn't mean it as a punch line. That's what he does: customer service. And that is the essence of craigslist's worth. If the community becomes overrun with spammers and scammers, it will lose value for its members. So Newmark listens to their complaints and fixes problems. Craig is the cop.

When Newmark spoke to my students at the City University of New York Graduate School of Journalism about some of the projects he was engaged in for the public good—such as investing in the future of quality journalism—one of them asked why he didn't sell craigslist for billions of dollars, which he certainly could, and turn his assets to philanthropy. Newmark said he believes he is helping people more by keeping money in their pockets and away from middlemen. He attributes the success of craigslist to treating his community as stakeholders, and he is paying them their internet dividend.

Newmark operates by many of the rules in this book. He created a platform and network for his communities. He trusts the wisdom of his crowd. He brings communities elegant organization. He understands that free is a business model. He relies on the gift economy. He dooms middlemen. He runs a disarmingly simple system. But then he adds his own unifying principle of technology, communities, and the internet. Here it is, with classic Craig brevity:

"Get out of the way."

That's it, Craig's Law: Get out of the way. If you make a great platform that people really want to use, he argues, then the worst thing you could do is to put yourself in the middle, getting in the way of what people want to do with it. As a customer, I often feel that airlines, cable companies, phone companies, insurance companies, doctor's offices, car dealers, banks, schools, and government agencies exist to get in my way—it's their business model. Not Newmark.

When he started craigslist as an email list, Newmark will tell you he had no idea what it would turn into. He didn't know the impact it would have on classifieds and news. He didn't know that people would find each other and go on dates and find restaurants for those dates and get married and have kids and buy baby furniture and get apartments and buy cars and improve their lives thanks to his simple lines of code. But they did. He didn't know that when Hurricane Katrina struck, the New Orleans diaspora would use craigslist to find each other and jobs and homes. If he had tried to anticipate that, if he had over-researched and over-designed and set up all kinds of rules, restrictions, navigation, instructions, and fees for how to use craigslist, Katrina's people wouldn't have done what they did. And craigslist would be smaller than it is.

Instead Newmark created something useful that people used. He stood back and let them do it. He listened to them and added the features they wanted. He kept listening and solved problems with the technology and with the community's use of it. And, by the way, his is about the ugliest but most useful design you can find this side of Google.

Google, too, tries to get out of the way. It creates platforms that people use—even enabling them to build businesses atop them—in ways that Google could not predict, could not design around, does not limit (well, not much), and generally does not charge for. Google realizes that its real value is not in limiting what people can do but in helping them do what only they can imagine. That is the essence of the Google worldview. That is what I will try to apply to a host of companies, industries, and institutions in the next section of this book. We end this section with the single best bit of advice you can glean from Google and from craigslist: Make something useful. Help people use it. And then . . .

Get out of the way.

If Google Ruled the World

"The search engine is going to control the planet," declared author Paulo Coelho. But surely not everything, right? It's not as if Google would want to run something dull like a utility (except that it is investing in the power industry) or a telephone company (well, it almost has) or enter the health industry (but it just did) or open a restaurant (then again, its cafeteria is world-famous and so is its chef, who wrote the book *Food 2.0*). Some people wish Google would take over a newspaper—The New York Times is often nominated—or entertainment companies or perhaps the software giant Microsoft. But no, Google knows what it is. Its ambition is not to take over the world, but to organize it.

So now that we have distilled Google's success into a series of laws and lessons, we will attempt to apply them to a number of industries and institutions. I won't pretend that I can fix a company in just a few pages. If only it were that easy. . . . Nor will I claim that I have found all the secrets to Google's success. If only I could. . . .

The point, instead, is to see things differently, to understand the fundamental changes of the Google age, to ask hard questions, to grasp new opportunities, to rethink, reimagine, and reinvent. That is the example to follow from Google. So the specifics of these cases are less important than the discipline, the attitude, the imagination, and the courage it takes to lead in this era of magnificent upheaval. Even if you don't work in, say, the ad industry, I hope that in the discussion of how to remake advertising, you will find ideas and inspiration for your own situation. These industries and institutions provide a wide variety of examples of how to live by Google's rules. Not all the rules will apply to your particular circumstances. But thinking and seeing in new ways is an imperative for everyone.

"Google has brought massive shifts in the way different generations and people think," said ad industry visionary Rishad Tobaccowala. He predicted that as a result, "there's going to be a huge new business which is built around the psychology of being analog in a digital world—everything that has to do with therapy. Which I think is why spas are so big now." That may be why Google's offices feature pods where employees can shut off the world and easy chairs where they can just stare at an aquarium. It's stressful organizing and changing the world. But before you check yourself into the sanitarium, remember that the real moral to the Google story is this: If Google can do it, so can you. Google is seeing problems, solving them, and finding opportunities in them by thinking in new ways. This is all about finding your own new worldview.

There are two ways to attack the problems of these industries: to reform the incumbents or to destroy them. In some cases, we'll take one path, in others both. But in any case, the wise course is to destroy your own models before some kid in a garage—or in a Harvard or Stanford dorm room—figures out a way to do it for you. Think like Google, succeed like Google—before Google does.

Media

The Google Times
Googlewood
GoogleCollins

The Google Times: Newspapers, post-paper

On what turned out to be an eventful week in London in 2008, Edward Roussel, digital editor for the Telegraph Media Group, told me over tea and toast that he had pondered the question I ask in this book's title. He answered it with a striking vision for newspapers: What if papers handed over much of their work to Google? Roussel reasoned that Google already is their best distributor online. He couldn't imagine a paper creating better technology or attracting better technical talent than Google. In advertising, Google is the clear winner. So why not outsource distribution, technology, and a good share of ad sales to Google as a platform so the paper could concentrate on its real job—journalism?

Roussel was following a key rule in this book: *Decide what business you're in.* The next day, I issued the same challenge to his competition, the Guardian, where I work and where I wound up a series of seminars on the future of journalism. My assignment was to pose 10 questions papers should answer now. The first: Who are we? Papers must no longer think of themselves as manufacturers or distributors. Are they in the information business? That would seem obvious, but when information can be so quickly and easily commodified, it is a perilous position. Are they in the community business, like Facebook? Not quite; few papers enable communities to organize themselves. Are they in the knowledge business, like Google or Amazon? Not yet; they haven't put themselves in the position to know what their readers know. In the end, I urged papers to become platforms for larger networks of news—but they're not there yet.

The night before, the Guardian had invited Arianna Huffington, founder of The Huffington Post, to speak. She announced she was taking her service local and would invade Chicago, hiring one editor to build a site around the best bloggers there. A reporter at the beleaguered Chicago Tribune—which now had Huffington's target on its back—asked me how the paper should respond. My reply: The old way would have been to treat Huffington as a competitor. The new way would be to find the means to work with her: Sell local ads for her and get a piece of her revenue. Quote her bloggers in the paper, taking advantage of her recruiting and relationships and earning friendship—and links—in return. Start new blogs that Huffington's writers would want to talk about and link to. Give Huffington headlines from the Tribune, which also link to the paper. The Tribune no longer owns the market, I told him. Its ambition should be to join and help a network.

News organizations don't yet think that way. That same week, while in London, I became embroiled online in a bloggers' battle with the Associated Press, which had sent legal letters to a site demanding that it take down excerpts of its stories, some as short as 33 words. The AP thought the bloggers were stealing its words. Bloggers, however, believed they were doing the AP a favor every time they quoted and linked to its stories.

In this confrontation, we witness the millennial clash of old and new media models: the content economy vs. the link economy. The AP, like the papers it serves, thought its content was its value and its magnet. But online, content without links is the tree that falls in the forest that nobody hears (and turns into newsprint). So the real value in this transaction was not the content that was, in the AP's view, stolen, but the links that were, in the bloggers' view, given. The content economy made money by controlling and selling content. In the link economy, it no longer pays to sell copies of content when the original is just a link and a click away.

This link economy makes five demands: First, you must produce unique content with clear value; commodity content will get you no links or Googlejuice. Second, you must open up so Google and the world can find your content. (*If you're not searchable, you won't be found.*) Third, when you get links and audience, it is up to you to exploit them, usually through advertising. Fourth, you should use links to find new efficiencies. (*Do what you do best and link to the rest.*) Fifth, find opportunities to create value atop this link layer: curation of the best content; technology infrastructure to enable links to be found; and advertising networks to help creators monetize links and traffic. Exploiting this sort of tectonic shift—seeing how the world is disrupted and finding opportunity in it—is a key skill of Googlethink.

For news organizations, going digital is not as simple as filling web pages. This transformation requires them to reinvent themselves—how they think of themselves, how they operate, how they relate to the public, how they make money—and fast. Jeffrey Cole of the University of Southern California Annenberg School's Center for the Digital Future found in a 2007 survey that young people 12 to 25 will "never read a newspaper." *Never.* Philip Meyer wrote in his 2004 book *The Vanishing Newspaper* that if current trends continue, the last American paper will be published in 2040—and that downward slope has only steepened its decline since he said that. This is not a drill.

Google's impact is more direct and immediate on media than on other industries—though their turns are coming. So as a demonstration of the discipline of adapting Google's rules, I begin this chapter—unlike others to follow—by laying out relevant rules and explicitly interpreting them for newspapers.

Atoms are a drag. Newspapers assumed their competitive advantage was in owning the means of mass production and distribution. In the old, scarcity-based content economy, they were right. But now print's infrastructure carries an unbearable cost burden. So I say papers should set a date in the not-too-distant future when they will turn off the presses.

Foolish, you say? Old mass media still has value, you argue. Online revenue is not meeting print revenue. As readers move to the internet, newsstand money disappears. In advertising, print dollars are replaced by mere online dimes. Don't they still need paper? Yes, but the scale of newspapers' businesses will never be the same now that they no longer hold local monopolies. In the shift from physical to digital and mass to niche, the best way to exploit the legacy value of a paper is to use its old-media megaphone to promote and build what comes next. First, a paper has to decide what is next. It has to design and build its post-paper products—retraining and restructuring staff and sloughing off unnecessary costs—before the presses go silent. It has to promote the new products even at the expense of the old: Cannibalize thyself. Convincing audience and advertisers to move to the future is better than following them there after they have discovered other sources of news.

Casting off atoms will allow newspapers to brag: no more dead trees and lost oxygen (an ecological site calculated that newsprint production used up the equivalent of 453 million trees in 2001); no more gas-sucking, pollution-spewing trucks to haul them around; no more presses draining energy; no more waste to recycle; no more oil pumped to make ink. To hell with going carbon-neutral. A former paper is an ecological hero!

In 2005, just after it had finished installing new, smaller-format presses at a cost of $150 million, the Guardian invited me to talk with its managers about what would come next—digital. Editor Alan Rusbridger stole my thunder when he conceded that those presses were likely the last they would ever buy. "The last presses." I couldn't imagine an American publisher saying those words except with his dying breath. Rusbridger saw it as his job to deliver the Guardian over the chasm it faced from print to online, atom to bit. His mission wasn't to shelter the old medium but to take its values to the new world as quickly, safely, and sensibly as he could.

Paper may not disappear. But if newspapers do not at least plan for the eventuality—if not inevitability—of the transition, they will be left protecting nothing but their presses. Again, protection is no strategy for the future.

Think distributed. News organizations can no longer rely on the idea that the world will beat a path to their door. People are finding their own ways to news through no end of new routes: friends' blogs, aggregators such as Google News and Daylife, collaborative news sites such as Digg, feeds on Facebook or Twitter, apps on mobile phones, and who knows what comes next. As a college student said in The New York Times in 2008: "If the news is that important, it will find me." Thus news organizations should stop presenting themselves as destinations and start seeing themselves as services, pushing out feeds, offering content to networks of sites, getting their news to where the people are. This is the new home delivery, the internet as paperboy.

Be a platform. Join a network. You can't do it all yourself anymore. By joining collaborative networks, you can get help. For newspapers, that may mean soliciting the public's assistance in finishing stories. It may mean recruiting and mobilizing the public to report. It may mean setting them up in business. It certainly means welcoming their contributions and corrections (one way to follow the rule, *make mistakes well*).

Newspapers can provide collaborators with raw material to create products—news reports to comment on, video to remix, assignments to follow. The New York Times and NPR each announced programs to make content available for mashups and remixing via APIs (application programming interfaces). Newspapers can also provide functionality—blogging tools and the means to repackage, say, Google Maps into collaborative community resources. They can educate collaborators, sharing what they know about how to get access to public information, avoid libel suits, or shoot video (as the Travel Channel and some local TV stations do). They can give good sites promotion and traffic. They can generate revenue by setting up ad networks for these collaborators, following Glam's example. The papers, in turn, get news and information they couldn't afford to gather on their own at lower cost and with lower risk, and they become part of something bigger than themselves.

Or that's the theory. A holy grail of online newspapers—as yet unattained—has been the idea of collaborative hyperlocal news networks: armies of blogging neighbors who gather and share news and photos from their school boards and street fairs. There have been many attempts to reach this goal and about as many failures, no shortage of them mine. I learned that it was a mistake to expect people to come to my newspaper site and contribute their work; often they want to own their own stuff in their own space. I also learned that bloggers need the means to support what they do—that is, money.

In 2004, I held a Meetup to persuade people to blog on NJ.com. Good idea, said journalist Debra Galant, but it's too good an idea to do it for your site, Jeff. She started her own blog, Baristanet.com, which covers Montclair, New Jersey, and now serves 10,000 daily readers and 100 advertisers. What should its relationship be to the site and paper I worked with, The Star-Ledger? Rather than competing, they collaborated in 2008 to print a joint guide to Montclair, sharing content and credit, with both the paper and the blog selling ads. It's a start. Next, I'd like to see a network of scores of Baristanets covering hundreds of towns and eventually thousands of interests.

Collaborate. Collaboration is co-creation. It requires giving up some control of assets so collaborators may remix, add to, and distribute content. The newspaper gets more content and gets talked about, which is how it will get new links, readers, attention, loyalty, and Googlejuice.

In 2007, Brian Lehrer's public-radio show on WNYC wanted to use its ability to mobilize the public for a project in collaborative journalism. Lehrer asked his listeners to go to their local stores and report the price of milk, lettuce, and beer. Hundreds did, giving the station data no single reporter could have gathered alone. WNYC plotted the data on Google Maps, showing which neighborhoods were being gouged. It also learned that some stores were charging illegally high prices for milk.

The BBC opened up many of its resources in a public laboratory called Backstage, which enables anyone to build products on top of its content and data. Remixes have included a service that took BBC news feeds and searched for related material from citizens on YouTube and Flickr; a service that found out which BBC stories were the most talked-about on the web; and one that mashed up road-traffic data atop Google Maps. The BBC—like Facebook—attracted scores of developers making new products that made the BBC more useful and brought new ideas to the media giant without the cost or delays huge organizations bring. *Welcome to the open-source, gift economy.*

Listen well. Just as About.com and Google monitor search requests to see what the public wants to know, so newspapers should create the means for the public to say what it needs to know and to assign work to journalists. BusinessWeek is soliciting such requests. Digg.com had its users vote on questions it would ask politicians at the 2008 political conventions. In 2007, I worked with trainees at the German publisher Burda, brainstorming products. One of them asked a question so obvious I kicked myself for never having asked it myself: "Why doesn't the public assign us?" Right. Readers know what they want to know. Journalists need a means, like MyStarbucksIdea, to gather assignments. This mechanism turns the relationship between the journalist and the public on its head. The public is now the boss. If journalists are uncomfortable with that, it means they don't trust the public they serve. Remember: *Your crowd is wise.* Remember, too, Weinberger's Corollary: *There is an inverse relationship between control and trust.*

The internet kills inefficiency. Newspapers are inefficient enterprises—because, as once-rich monopolies, they could be. When Rupert Murdoch acquired The Wall Street Journal, he complained that 8.25 editors touched each story. At The New York Times, there are three editors for every writer. When Sam Zell took over Tribune Company, he had efficiency experts count how many inches of text writers produced. These may be shallow metrics, but they reveal much room to change. And that change is coming as, according to the blog Papercuts, newspapers laid off 12,299 journalists in the first 10 months of 2008. Once a paper decides what it is, it's clear that it must marshall all its forces behind one goal. For local papers, that should be local reporting.

The mass market is dead. Long live the mass of niches. Papers should no longer make just one mass product, a newspaper. Some are producing new services for more targeted interests, locales, and communities: hyperlocal sites and papers; a local sports talk show; a local golf magazine; a mobile weather service; local job fairs; parents' guides. These products need not be created and owned by the company; they can be produced by others and distributed or sold by the paper. The more communities served, the better. *Small is the new big.*

Elegant organization. A paper should provide its community with what Facebook's Mark Zuckerberg gave his. In a sense, papers always have. They organize a community's knowledge so it can better organize itself. Now there are more tools to do that. Papers can create platforms where neighborhoods, towns, schools, clubs, or people with like interests can share what they know and editors can bubble up news out of that. Once the platform has been created, they should follow Craig Newmark's advice: *Get out of the way.*

Beware the cash cow in the coal mine. Papers sat back on their cash flow and assumed something would rescue them. Nothing did. Now papers will die. But the demand for news will not go away; it's growing. New products and competitors will emerge and there'll be enough audience and money to support them—if they are not saddled with the costs of printing. Can the papers that survive invent these new products themselves with their cultures? Jim Louderback, CEO of the internet TV company Revision3 (more from him next) has this advice for legacy companies: "Look at how Steve Jobs made the Mac. He took a core group of people and put them in a closet somewhere and they built something completely different. So take a core group of people and put them in Kentucky or St. Louis and build something entirely new." Rethink everything: What is a news story? Is a topic page a better vehicle for covering local news? How should news be gathered? How should it be shared? How should it be supported? *Encourage, enable, and protect innovation.*

What does a newspaper look like if it is no longer a newspaper? It will be more of a network with a smaller staff of reporters and editors still providing essential news and recouping value for that. Paper 2.0 will work with and support collections of bloggers, entrepreneurs, citizens, and communities that gather and share news. A newspaper is no longer a printing press that turns out money. But as a network it could be bigger than papers have been in years, reaching deeper into communities, having more of an impact, and adding more value. To get there, it has to act small but think big and see the world differently.

Googlewood: Entertainment, opened up

Entertainment is built on a blockbuster economy: Hits are huge and everything else is merely the price you pay to play the odds. This system has long been fed by scarcity: only so many movie screens, so many hours of TV seen by only so many viewers, and so many shelves in the record store (when there still were records and stores to sell them). The audience was herded together to consume a limited field of choice, and the winners were the products that appealed to the most people. There will always be blockbusters just because some things are that good (great movies) or because we enjoy talking about shared experiences (silly reality shows) or because the hype is too huge to ignore (the Oscars). Hollywood is eternal.

The economics of abundance—the mass of niches, the long tail—has opened up entertainment's business models in ways we have not seen since the last waves of new media technologies: sound recording, film, broadcast. Today we can watch whatever we want. Hell, we can make whatever we want. It will become harder and harder to turn out blockbusters because there is so much more competition for our attention. But it also will be possible to produce more entertainment more people like—that is our new abundance.

Hollywood was built on a system of control. You could break in only if you made it through a gauntlet of agents, executives, and distributors who controlled the money and access to the audience. The internet is busting that system apart. But we didn't need to wait for the web to break free. It was possible to be a rebel before. It's just easier now.

I return to Howard Stern, who is not only the self-crowned king of all media but who was, I argue, Googley before there was a Google. He saw a radio industry built around the local broadcast tower and broke its rules, starting in 1986, when he built a syndicate of stations that made him famous (and infamous) across the country. He didn't rely on an existing network. He built his own network. Then he used radio as a platform to create a presence on TV. He used radio to become a best-selling author, and he turned his book into a hit movie. He later became huge on the internet, and put satellite radio in orbit.

Stern's relationship with his audience is what set him apart. He created a collaborative product—not just because he took phone calls from listeners but because those listeners made their own entertainment, which they generously gave to the show: phony phone calls, brilliant song parodies, theme songs for hapless producer Gary "Baba Booey" Dell'Abate, games, even movies. They gave him their creativity and loyalty. He gave them airtime and attention. This was their mutual gift economy.

Stern decided long ago that he would not push a self-serving charity as rival Don Imus had or sell tacky schwag like the Rush Limbaugh Excellence in Broadcasting mouse pad. I wouldn't mind buying a Stern hat or jacket—I'd wear my taste proudly—but Stern won't sell them to me. He refuses to cash in on our relationship. He knows that his value rests with his fans. Stern took a gamble on that relationship in 2006 when he moved from broadcast—chased off by the Federal Communications Commission's harassment—to Sirius Satellite Radio. He received a reported $500 million for the move—motive enough, of course—but there was no way to be sure that the millions of fans needed to make him worth his price would follow. They did. At Sirius, Stern has handed over control to his audience; when they told him to change programming on his two 24-hour satellite channels, he obeyed.

I use Stern as a case study in Googlethink to demonstrate that you don't need to be Google—or be on the internet or rely on technology or even be inspired by Google—to think in these new and open ways. Stern broke the control system and rules that the entertainment business holds dear and built his empire on his relationships. It's still about relationships.

The internet just makes it easier to break rules and break in. Anybody who's any good can aspire to be a monarch of any or many media. They may not be as big as Stern, Jon Stewart, or Steven Spielberg. But in a post-blockbuster, small-is-the-new-big economy, they don't have to be.

Now fast-forward to 2005, when geek-show host Kevin Rose left TechTV after his network merged with G4, a game channel. Instead of getting another job at another network, Rose started his own networks, because he could. First he created Digg, a collaborative news service where users suggest stories and then vote on them to create the community's front page. It attracts more than 25 million users a month. The service was revolutionary, giving the public—rather than editors—the power to make news judgments. Of course, the public always had made its own judgments; Rose just recognized that and enabled them to do it together.

Then Rose started his video network, Revision3, and the first show on it, *Diggnation*, in which he and his former TechTV colleague Alex Albrecht sit on a grungy couch with a different beer in hand each week talking about some of Digg's favorite stories for more than 30 minutes straight. If one of them has to do what one must do after drinking beer, they don't stop the tape; Alex just gets up and goes to the bathroom. The show could not be more casual and less like TV, but that is precisely its authority. My son, Jake, is a fan—he introduced me to it—and I tried to repay the favor by sharing professional podcasts about technology from NPR and the BBC. As soon as I played them, I realized they didn't hold the same authority as Digg because they were too packaged and plastic.

Diggnation draws an audience of 250,000 each week (a nighttime cable news show on some networks is satisfied with 150,000 viewers). Just because it's on the internet doesn't mean it's small. But its costs are. Nonfiction TV—news shows, not scripted dramas—on broadcast networks costs about $300,000 an hour to produce. An hour of Revision3 programming costs a tenth of that. Internet TV can get even cheaper. In 2007, I visited 18 Doughty Street, a five-hour-a-night Tory talk-show network in London that broadcast on the internet from a townhouse living-room set with all the trappings of TV: couches, seven cameras, a control studio, and potted palms. I asked Iain Dale, the founder, to calculate his per-hour cost for talk. It came out to $140. Sure, the comparisons are unfair. News networks have journalists, bureaus, producers, executives, expensive anchors, writers, makeup people, hair people, camera people, sound people, directors, and free muffins. But do they need all that? In 2007, I wrote and recorded an opinion piece for a short-lived segment on the *CBS Evening News* (it never aired—my mention of dethroned anchor Dan Rather may have had something to do with that). Up to the taping, I saw 12 people involved, which didn't include countless more editors and unseen executive producers and technicians. That day at home I used the same script to record the same opinions on my Mac. Cost: zip.

Movies are worse. Not long ago, I happened on a studio shoot in Manhattan. Even though I'd covered the industry for years, I was amazed anew at the cost of it, at the stuff they drag around. On one truck, a huge container was filled with nothing but blocks of wood with Paramount's logo burned onto them. Of course, studios do need much of this stuff to make movies that will look great on a big screen. But do they need it all? *Diggnation* has just a camera pointed at its couch. It entertains, too.

In the text web, the delta—the difference—in cost between the old and new way is enormous; that is what has led no end of bloggers and newcomers to create content sites. In film and video, that delta is many times larger, which I believe will lead to even more investment in online shows, as the opportunities are even greater. Revision3 started on a shoestring but received a reported $9 million investment to create more shows, build a studio, and hire its CEO. It's still run on a shoestring, CEO Jim Louderback told me. "The story of the internet," he said, "is ruthlessly efficient business models and blowing away barriers to entry and access."

Revision3 saved money on equipment, which Louderback credited to Moore's Law. Intel's Gordon Moore decreed in 1965 that the number of transistors and thus the computing power on a chip would double every two years (this law enabled Google and the internet to exist and led to every law in this book). The cost of digital cameras has thus plummeted. Revision3 goes Cadillac with an $8,500 model but I've seen newspapers and even TV stations recording high-definition segments with $1,000 handhelds. Instead of a fancy TelePrompTer (and expensive writing team to fill it with words), Revision3 uses a cheap LCD screen and mirror. Instead of editing suites that once cost tens, even hundreds of thousands of dollars, they edit on Macs. The only equipment that doesn't benefit from Moore's Law, Louderback said, is the handmade Italian pedestal for moving cameras while shooting. It has no electronics but relies on precision ball bearings. Damned atoms.

Staff costs are low, too. Instead of hiring pretty faces with good hair to read the words writers put on TelePrompTers, Revision3 hires hosts with knowledge and passion about their topics and the ability to attract a community. Distribution costs little because there are so many partners, including Google's YouTube, that can spread video around. Marketing? No need for that when you have a loyal audience. I stood in that audience when *Diggnation* came to New York and 2,000 people showed up (I was the oldest geek there and sympathized with my son, who was standing next to the only head of gray hair in the place; it was like having your mom take you to a Stones concert). To market itself, Revision3 cuts up its shows and puts the best bits on YouTube so fans can pass them around—a demonstration that your product can be your ad and your customers are your ad agency.

What about revenue? Louderback said that by the middle of 2008, a show the size of *Diggnation* was selling three sponsorships per episode at a cost of $80–$100 per thousand viewers (the standard measurement for advertising). By contrast, banner ads on web sites can sell for as little as a few dollars or even cents per thousand. How can *Diggnation* command that premium? Once more: relationships. The hosts deliver the commercials and viewers remember them. Louderback said 100 percent of audience members could recall the name of one sponsor without help and 93 percent could name two. That is unheard of on television, where commercials are ignored or skipped. So do the math: With an audience of 250,000 per week, that could work out to as much as $4 million a year and growing. Not bad for two guys on a couch.

Revision3 moved past tech to shows on magic and comic books. Louderback finds talent not on TV but online inviting viewers to submit their own pilots. The internet is an amazing source of new voices if you know how to listen for them. Talent may not be everywhere, but it's not as scarce as once thought.

The key, Louderback said, is to realize that the internet "is a new medium. It's completely different. Think how Ted Turner created CNN. He didn't just think about plopping a broadcast network on cable. He thought about creating an entirely new medium." So did Kevin Rose. His shows are communities. He is the new Turner, Murdoch, Hearst—or Oprah. He is the next-generation media mogul because he thinks differently.

This new relationship we have with—in the words of New York University journalism professor Jay Rosen—"the people formerly known as the audience" is collaborative. I don't mean we'll each end up picking our own endings to a movie. I don't want that. Writing the ending is the job of the author. Still, entertainment is becoming collaborative. When LonelyGirl15—the saga of a pretty teen girl talking about her odd life via a webcam in her bedroom—became an entertainment phenom on You-Tube, what was most fascinating was not the LonelyGirl videos but the videos viewers made around them, responding to her, asking questions, affecting the course of the narrative. When it turned out that LonelyGirl was not real but an act of fiction, the audience's videos—many exhibiting anger and disappointment—were captivating. The art was the collection of everyone's work, creators and audience. The art was interactive. Something similar happens on discussion forums such as Television Without Pity, where producers take advice about plotlines and characters in series that threaten to jump the shark. These producers realize that the audience owns a show as much as its creators.

Entertainment can now break out of its old forms. Comedy doesn't need to be 22 minutes long (plus eight minutes of ads). Movies can become serials. Shows can be collaborative. Talent can come from anywhere. Audiences are distributors. We can watch entertainment anywhere.

Hollywood—particularly TV—has not been blind to this change and learned from the music industry as it imploded trying to maintain control in an uncontrollable world. TV networks might just save themselves because they broke their own rules. ABC was willing to hurt its distributors—local stations—when it streamed shows on the internet and sold them on iTunes. NBC and Fox created an impressive player called Hulu; in the U.K., the BBC started its equivalent in the popular iPlayer. Like Google, they learned to think distributed.

What will Hollywood studios and TV networks look like in the Google age? At one level, they won't change: They will still pray for blockbusters and the stars that make them. At the top, the celebrity economy is largely immutable because there can be only so many big stars at once. But from the bottom, we will see more, if smaller, celebrities in many variations on Warhol's Law: Everyone is famous for 15 clicks, links, tweets, or You-Tubes. Fame, like talent and audience, is no longer scarce.

Managing this abundance presents many opportunities. More than ever we need guides. Too bad TV Guide is choking in the coal mine.

One-size-fits-all criticism won't work anymore. But a system that helps us help each other find the best entertainment would be valuable. If I were to start Entertainment Weekly today, it would be that: a way to find just what I like, a collaborative Google of taste.

Entertainment will be more of a social experience. Though I still want authors to do their duty and polish stories, that doesn't mean I wouldn't like to see other people remix shows and movies. In the old, controlled way of thinking, remixing was a violation of copyright. In the new, open, distributed model, it is how you join the conversation. Comedy Central's Stephen Colbert has—Stern-like—challenged his audience to remake videos of him and of John McCain. Some were great, some were nowhere near that, but in the process, they spread his challenge all over YouTube, MySpace, and blogs. It's a gift economy and it's an ego economy: Everybody who made a video wanted attention and could get it from Colbert and his community. The content was the advertisement, viewers were the creators and distributors, and Colbert was the catalyst. Maybe that's what entertainment becomes: the spark that inspires more creativity and attracts not just audiences but communities of creation in a million Hollywoods.

GoogleCollins: Killing the book to save it

I confess: I'm a hypocrite. If I had followed my own rules—if I had eaten my own dog food—you wouldn't be reading this book right now, at least not as a book. You'd be reading it online, for free, having discovered it via links and search. You'd be able to correct me, and I'd be able to update the book with the latest amazing stats about Google. We could join in conversations around the ideas here. This project would be even more collaborative than it already is, thanks to the help of readers on my blog. We might form a group of Googlethinkers on Facebook and you'd be able to offer more experience, better advice, and newer ways to look at the world than I alone can here. I wouldn't have a publisher's advance but I might make money from speaking and consulting.

But I did make money from a publisher's advance. That is why you are reading this as a book. Sorry. Dog's gotta eat.

I already do most everything I describe above, not in this book but on my blog, where ideas are searchable and collaborative and can be updated and corrected—and where I hope conversations sparked by this book will continue. I believe the two forms will come together—that's part of what this chapter is about. In the meantime, I'm no fool; I couldn't pass up a nice check from my publisher, Collins, and many services, including editing, design, publicity, sales, relationships with bookstores, a speaker's bureau, and online help. There's a reason publishing is still publishing: It still pays. How long can it stay that way? How long should it stay that way?

As I suggested that papers should turn off their presses, I have a suggestion for book publishing: We have to kill books to save them. The problem with books is that we love them too much. We put books on a pedestal, treating them as the highest form of culture: objects of worship, sacrosanct and untouchable. A book is like a British accent—anything said in it sounds smarter, even if it's not. But, of course, there are bad books. Any episode of *The Office*, *The Wire*, and *Weeds*, to name just a recent few, is better than too many books on the shelf. Yet we dismiss TV as our lowest cultural denominator, and we allow government to censor TV shows whereas we would not permit it to ban books. Books are holy.

We need to get over books. Only then can we reinvent them. Books aren't perfect. They are frozen in time without the means to be updated and corrected, except via new editions. They aren't searchable in print. They create a one-way relationship: Books teach readers, yes, but once written they tend not to teach authors. They cannot link to related knowledge, debate, and sources as the internet can. David Weinberger taught me in *Everything's Miscellaneous* that when knowledge is frozen on a page it can sit in only one place on a shelf under one address so there is only one way to get to it. In the internet age, with its many paths to knowledge, this, too, is a failing of books. Books are expensive to produce. They depend on scarce shelf space. They kill trees. They rely on the blockbuster economy, which is to say that only a few are winners and most are losers. They are subject to gatekeepers' taste and whims.

Books aren't read enough, I think we'd agree. Don Poynter at BookStatistics.com compiles sobering stats about the industry and reading. Citing BookPublishing.com, he reports that 80 percent of U.S. families do not buy or read a book in a year; 70 percent of U.S. adults had not been in a bookstore in five years; 58 percent of U.S. adults don't read a book after high school (though this conflicts with National Endowment for the Arts stats saying that in 2004, 56.5 percent of U.S. adults said—said—they had read a book in a year). Books are thrown out when there's no space for them and end up as trash or pulp. Forty percent of books that are printed are never sold. Books are where words go to die.

When books are digital, all kinds of benefits accrue. Books can become multimedia, like Harry Potter newspapers, with moving pictures, sound, and interaction. They can be searched, linked, and updated. They can live forever and find new audiences anywhere. Conversations can grow around ideas in books, exposing them to new readers. Writing in Library Journal, Ben Vershbow of the Institute for the Future of the Book envisioned a digital ecology in which "parts of books will reference parts of other books. Books will be woven together out of components in remote databases and servers." Kevin Kelly wrote in The New York Times Magazine: "In the new world of books, every bit informs another; every page reads all the other pages." When an idea is spread among people, it can grow and adapt and live on past the page. Before a convention of booksellers in 2006, author John Updike called Kelly's vision of "relationships, links, connection and sharing" Marxist and "a pretty grisly scenario."

There's just one problem with these visions of digital publishing paradise (including mine): money. How will authors be paid for going to the trouble of reporting, imagining, and writing when so much of that is free on the internet? The internet is unsympathetic.

Robert Miller, former publisher of Disney's Hyperion, came to HarperCollins—parent of my publisher—as I was writing this book. His mission was to update the business of book publishing and its two dogging problems: advances and returns. The difficulty, he explained to me, is in the middle. At the top, best sellers make money and at the bottom, we now have the means for no end of niches to create small books (six huge publishing conglomerates control the high-end of the market but Publishers Weekly reports that the total number of publishers grew from 357 in 1947 to 85,000 in 2004; that's a lot of niches). In the middle, however, advances to authors (like me) have been rising, increasing risk and losses.

It's a problem of the blockbuster economy: Publishers throw a lot at the wall, hoping something will stick but never knowing what will. Though ownership of publishing houses has consolidated, Miller said that hasn't much affected competitive bidding among them. All it takes to pump up the price is for two houses to want the same book. That has been the case since 1952, when literary agent Scott Meredith started auctions among publishers rather than sending a book to one house at a time, as the gentlemen of the trade used to do. Today most books don't earn enough to pay for the advance publishers give authors. Miller said a house is doing well if 20 percent of books earn back their advances. Imagine any other industry in which 80 percent of the products you produce lose money. It's a growing insanity.

Miller's proposed solution: He is offering smaller advances—maximum about $100,000—and in return, authors split a book's profit, 50-50, with the publisher (for comparison, I receive a 10–15 percent commission of the retail price in hardcover and 7.5 percent in paperback and we split fees from international sales). The idea is that author and publisher share the risk and the reward.

Then there is the problem of returns. Publishing is a consignment business. Bookstores can send unsold books back to publishers—a practice for which Simon & Schuster gets the blame—so it's the publishers who bear the risk, not to mention the huge cost of printing, shipping, storing, and pulping all those unwanted books. Books are atoms of perishable value. Miller wants to offer booksellers, too, a higher cut of profits if they will take the risk of owning the books they order. The resultant risk to publisher and author could be that bookstores won't order enough to meet demand, but Miller said publishers are increasingly good at printing more copies quickly.

Miller's goal is to make the existing print business more profitable. That's fine as far as it goes. He acknowledges that there are other models that need to be tried. Perhaps you could buy a book chapter by chapter as a Dickensian subscription: Buy enough chapters and you've bought the book (if it's bad, stop and you've spent less; BookPublishing.com says 57 percent of new books are not read to completion). Or buy the book in print and get access to it as an audio book and on an e-reader such as Amazon's Kindle. Some hold high hopes for print-on-demand, which would enable a store to sell you any book quickly, beating Amazon's delivery delays. But that's still expensive and it produces only paperbacks. Still, we know that readers will pay a premium for immediate gratification; that's why they still go to stores. Perhaps publishers could offer their own discounts if you're willing to wait a week or two, enabling them to collect orders until there are enough to print. They could charge less, still, if the reader is willing to take a book in the clumsy PDF format, which enables publishers to sell books to readers with no manufacturing cost. Or perhaps readers could subscribe to an author or series, guaranteeing the publisher and writer cash flow and a reason to publish the next book. Maybe authors could even tell readers that they'll write a book only if so many readers buy it in advance.

Peter Osnos, another publishing visionary on a mission to save the business, founded the Caravan Project to enable publishers to sell books in any form: in their traditional format, via print-on-demand, digitally in full or by chapter, and in audio. "When a reader asks for a book, the seller's answer should always be, 'how do you want it?'" he wrote at The Century Foundation. Osnos told me that the fundamental problems for publishing are availability and inventory management. If he can drive 20 percent of book selling to on-demand and digital, he believes he will save so much in printing unsold copies that he will be able to afford the marketing needed to make the business model work. He read a quote from The New York Times on the day that Google introduced its new Chrome browser arguing that Google needed to control its own destiny. That is the sense in which publishers should do what Google does, he said: control their own destiny.

Rick Smolan—best known for producing *America 24/7,* which chronicled one week in the life of the United States with 1,000 top photojournalists—has found another way to support his gorgeous and expensive photography books: sponsorship. "Why?" Smolan asked and then explained: "Because no publisher would publish our first book, *A Day in the Life of Australia*, we went to the business community in Australia and self-published the book—it went on to become the No. 1 book in Australia and sold 200,000 copies (in a market where 10,000 was a best seller)." More recently, he produced *America at Home* and a U.K. counterpart, each underwritten by an obvious sponsor—Ikea, which took little credit. (Smolan had another innovative idea: Readers can pay to get either book with their own photo on the cover.)

Why shouldn't books have ads to support them as TV, newspapers, magazines, radio, and web sites do? Ads in books would be less irritating than commercials interrupting shows or banners blinking at you on a web page. Would it be any more corrupting to have ads in this book than next to a story I write in BusinessWeek? You'd have to tell me. If I had a sponsor or two for this book, what would you think of my work as a result? If Dell bought an ad—because, after all, I do say nice things about them now—would you wonder whether I'd sold out to them? I'd fear you'd think that. What about a Google ad? Obviously, that wouldn't work. Yahoo? Ha! Who might want to talk to you and associate themselves with the thinking in this book while also helping to support it? Would it affect your thinking if the sponsorship lowered the price of the book? From the publisher's perspective, that could lower risk and increase profits. From mine, it could mean the book costs less and so it sells more and its ideas get wider distribution. (Come to my blog and let's debate ads in the paperback. Maybe we'll auction off a few pages on eBay.)

All these models still ignore the internet's greatest challenge: free. Free is going to kill publishing the way it killed music, right? Maybe not. Maybe free can save publishing.

The Googliest author I know, who also happens to be one of the most monumentally successful authors alive, Paulo Coelho, has nothing against selling books. He has sold an astounding 100 million copies of his novels and he estimates that another 20 million have been printed without authorization in countries that flout copyright. Even so, Coelho believes in giving away his books online for free. He's a pirate.

Coelho learned the value of free in Russia, where a pirated translation of one of his books went online. His sales there jumped from 3,000 to 100,000 to 1 million in less than three years. "So I said this is probably because of the pirate edition," he told me in a conversation in his Paris apartment. "This happened in English, Norwegian, Japanese, and Serbian. Now when the book is released in hard copy, the sales are spectacular. There's confirmation that I was right." He believes this piracy has helped make him the most translated author alive.

The pirated versions helped him so much that Coelho started linking to them from his own web site. After bragging about his openness at the Burda DLD conference in Munich in 2008—where I met him—he got a call from Jane Friedman, who was then head of his publisher, HarperCollins (parent of my publisher). "I was scared to death to talk to her because I knew what was coming: a tempest. She said, 'I have a problem with you.'" Friedman had caught him in the act of self-piracy when she discovered that one of the supposedly unauthorized pirate versions to which he'd linked still had Coelho's own notes and corrections in it. "She said,

'Paulo, come on, don't shit me.'" He sheepishly confessed to pirating himself. But he also said that neither of them could afford to lose face by taking the editions down; there'd already been publicity about them. They compromised: Each month, one of his books could be read for free, in full, in a special online reader that doesn't allow a user to copy (or search for or link into) the text. It's a start.

As this book goes to press, HarperCollins and I have discussed many digital options, including using that reader to put the book online in full for a few weeks before it is published, serializing pieces of the book online for a limited time, putting up free PowerPoint and video versions of the book, and more. I'll report on what worked on my blog.

In Coelho's view, the free web has given him more than book sales. He loves writing in a different voice in his blog. "I think your language for your blog is totally different from your language in the Guardian, right?" he said to me as I interviewed him for a column there. "We have to adapt ourselves. I have a lot of fun doing this." When I first met him, he said his blog would not influence his books. But six months later, just as he finished his latest novel, *The Winner Stands Alone*, he said his readers had been helpful explaining fashion and the attraction of brands to him.

Coelho Twitters. He uses a small Flip Video camcorder to record video questions for his audience via Seesmic.com, a video conversation platform. Inspired by his wired and eager assistant, Paula Bracconot, Coelho asked his fans to take pictures of themselves reading his books for a virtual exhibition at the Frankfurt Book Fair in celebration of his 100 million mark. Hundreds posted their photos on Flickr. Coelho also began inviting readers to his parties. The first time, he said on his blog that he would ask the first few readers who expressed an interest to a party he was holding in a remote Spanish town. Responses came in from all over the world and he feared that they expected him to pay their airfare. But they paid their own way, flying from as far as Japan. He webcast a later event and 10,000 people showed up online for it.

Coelho asked his readers to make a movie of one of his novels, *The Witch of Portobello*. With *The Experimental Witch*, he invited fans to film the stories of each of the book's characters. If there were enough good submissions, he promised to hire an editor to make the final cut. He also found sponsors—HP and MySpace—to pay for the project. As entries came in, he sent me links to them. Some showed remarkable effort and talent.

Note the common thread—from collaborative news gathering to news remixes for the BBC to Howard Stern's listeners' song parodies to Lonely-Girl15 videos to Coelho's open-source movie: Creation itself is a community. BookPublishing.com says 81 percent of Americans believe they have a book in them. None of them will ever be Coelho and Coelho's books will always be his. But creativity inspires creativity and the internet enables us to turn that into a conversation. The moral of Coelho's story, like that of so many here: It's about relationships. What has the internet given him? "It gives me a lot of joy," he said. "Because you are alone when you are writing." But no more. His goal online is to find relationships with more readers and sell more books. Coelho still believes in print. He lovingly patted a 3-D book—a thick biography of his rich life—and talked about the form's perfection.

Publishers treat Google as an enemy for scanning books and making them searchable (though you can't read them all cover-to-cover at Google .com). Instead, publishers should embrace Google and the internet, for now via search and links more readers can discover authors and what they say and develop relationships and perhaps buy their books. Authors can reach the huge audience that never goes into a bookstore. Publishers and authors can find new ways to bring books into the conversation. Books can live longer and spread their messages wider. I don't have the answers to books' challenges. But I know we must be willing to reinvent the form. The internet won't destroy books. It will improve them. Take Coelho's advice to publishers and authors: "Don't be afraid."

Just as I was dotting the final i on this manuscript, Google announced that it would create the means for publishers and authors of out-of-print books to receive payment from readers who want access to the full text online (Google will keep 37 percent of the fees as commission). Google also may sell ads on pages with book content and share that revenue with publishers and authors. Sergey Brin told a Wall Street Journal blog that the payment system could be extended to video, music, and other media.

This offer came in the settlement of a suit brought by publishers and authors fighting Google's scanning of books—seven million to date—to make them searchable online. But it is far more than a sop to angry book people. In one fell swoop, Google altered the life cycle and economics of books and potentially answered some of their most pressing digital needs. Now books will be able to live past the remainder table and pulper. They'll be searchable. They'll find new audiences over greater time and distance. They'll make more money. Google is not the enemy of books. It is becoming their platform for the future.

Advertising

And now, a word from Google's sponsors

And now, a word from Google's sponsors

Earlier, I argued that marketers' ultimate goal should be to eliminate advertising by improving their products and relationships instead. Consumers should be so lucky. Media companies supported by advertising should pray this never happens.

Media's prayers will be answered. We'll always have advertising and ad agencies because companies will never reach the nirvana of creating perfect products that every customer loves and sells for them. Marketers will still want to introduce new products and to envelop what they sell in the smoke-and-mirrors of premium brands.

In a sense, Google has changed advertising more than any industry I cover here. Google is in the ad business. It revolutionized the ad economy, enabling marketers to pay for performance rather than space, time, and eyeballs. It invented new means of targeting ads, making them newly efficient. It opened up millions more places to put ads, ending media scarcity. It attracted countless new advertisers. It dominates not only search ads but now web banner ads, and it has started selling ads in print and broadcast.

Yet for all the upheaval Google has brought to the advertising economy, ad agencies remain largely unchanged. That's because agencies still control the money, and nobody wants to mess with the guy who has the credit card. But their Google immunity will expire.

Rishad Tobaccowala, chief innovation officer of Publicis Groupe Media, started Denuo, a think-tank and laboratory inside his company, in an effort to create the next-generation agency. Asked what Google teaches him for this task, he counted five lessons.

First: Focus on talent. "Google feels like it was invented yesterday and it's a 10-year-old company already," he said. "AOL's a grandfather." Agencies are supposed to be fresh and young but Tobaccowala said they act old thanks to the "death grip" of years-long relationships among executives and clients. "Google would have talent running the place versus tenure."

Second: Newness. "In the service business," Tobaccowala said, "you take the form of the people you work for. If you really want to change, you need to get a new breed of clients." Google did that by creating a marketplace to serve the long-tail advertiser before the behemoths, "people who didn't think about advertising, who had no agency." They brought no rules, so they played by Google's rules.

Third: Data. Advertisers love data almost as much as Google does. They think it tells them where to spend their money and the return on investment they get. For decades, advertisers accepted dubious measurements of magazine readership (which assume that every allegedly well-worn copy is passed around to large groups) and broadcast audiences (surely they can't believe the Nielsen ratings). Then along came the most measurable medium in history, the internet, where advertisers can learn more about customers than ever before.

Fourth: Make money through the side door. "Google—and Apple—make money by giving a key part of their business away for free and then making money on something else." Too often, companies think that everything they do has value they must capture, charge for, monetize, preserve, restrict, and protect. Instead, the real value may come from the side.

Fifth: To quote Google's own No. 1 rule, "Focus on the user and all else will follow." Australian ad executive Peter Biggs spoke for much of his industry when he told ABC Radio National's The Media Report: "It's a consumer-driven business, but they are not our most important audience. Our most important audience is our clients, and their brands." Tobaccowala says the opposite. "Our fixation should not be on our clients. It should be on the people our clients want to engage, sell, and interact with. We should be the champions of those people. That is where we are missing the boat."

I wonder whether focusing on the consumer instead of the client ends up usurping much of the job of the agency as we know it. Fixating on customers should be the job of everyone—everyone—in a company. In business, we've long said we're customer focused. But today you have to mean it or your customers will call your bluff. Focusing on customers can't be outsourced to agencies.

Agencies will resist change until the economics of the industry change. Because agencies make a cut of what they spend, they are motivated to spend more on ads rather than to replace ad dollars with more valuable relationships between brands and customers. So clients may be the first to evolve. Just as I tell newspapers to imagine a day when they stop the presses and book publishers to think past the book, so I advise marketers to imagine as an exercise firing the agency, canceling the ad budget, throwing out the ads, and starting over. What is your relationship with your customers then? Where should you put your money? Where should you spend your first ad dollar and why?

Start, of course, by investing in your product or service. Tobaccowala said no amount of advertising will make up for a bad product. "Stop this yelling and screaming about what's your Facebook strategy," he tells clients. "Make absolutely certain that you have a great product or service. Make absolutely certain you have great customer service. Those are the first two rules of so-called advertising in this world. If you don't have those, don't pay any money to anyone to do anything."

Then turn the relationship with the customer upside-down. First, invest in customer service, making it a goal to satisfy every single customer. Remember that your worst customer is your best friend. Second, invest effort in social tools that enable customers to tell you what you should be producing; hand over as much control to them as you can (I examine this idea from another perspective in the chapter on manufacturing). The goal must be to produce a product people love. All companies claim that customers love their brands. But I mean customers love your product so much they want to tell the world—that kind of love, Apple love. Third, hand over your brand to your customers—recognizing that they have always owned it. Don't tell them what your brand means. Ask them what it means.

Every product is great; every relationship is satisfying—shoot for nothing less. So now you are spending quality dollars and relationship dollars over advertising dollars. You have handed over control of the product and the brand and gotten out of the way. If you haven't gone out of business by now and convinced every boss, board member, analyst, reporter, and stockbroker that you've gone mad, then it probably worked.

Won't you still advertise? Ask yourself why. To interrupt and irritate random people? No. To convince customers that a bad product is good? No. To inch ahead of your competitor with the brute force of media spending? No. To get people watching Sunday morning shows to buy your stock? Please, no. Do you advertise to tell customers something they didn't know and need to know about your product, such as an improvement or a better deal? Well, OK. Tobaccowala defines advertising as "the economics of information" (the title of a 1961 essay by Nobel laureate and University of Chicago professor George J. Stigler). Advertising is supposed to tell us about a product or its price so we can save effort, time, and money in our search for it. The internet has made that much more efficient. If the customers' goal is to reduce their transaction cost—the effort to find the right product at the right price—then doesn't the internet itself replace advertising? Often, yes.

A 2007 economics honors thesis by Daniel A. Epstein compared the pricing of similar cars listed in expensive newspaper ads with cars listed for free in craigslist. His hypothesis was that sellers who pay to advertise would want to price cars lower to sell more quickly so they would spend less on advertising. His research proved his theory wrong. Newspaper advertisers marked up their cars over Kelly Blue Book by 0.423 percent whereas craigslisters marked up theirs by 0.042 percent—a fraction. You might see this differential alone as motive to buy ads: Advertise and you can charge more. But I chalk this case up to a temporarily imperfect market, assuming that sellers who advertised were cagey and knew to ask for more, whereas craigslisters may have been bad negotiators who didn't know they could get more. As Google and craigslist push the market toward openness and transparency with more information and greater pricing competition, that alone will push prices down. Epstein's hypothesis may one day come true: Advertisers won't be able to afford to advertise and stay competitive.

Of course, there are still people who don't know about your product, who won't know to search for it because it is new or they are uninformed. In the classic case for advertising, they also may not know they have the problem you solve. In 1919, the ad agency for the deodorant Odo-Ro-No invented the term "B.O." and the insecurity around it. "Advertising," said the trade journal Printers Ink, "helps to keep the masses dissatisfied with their mode of life, discontented with ugly things around them." For good or bad, there will still be a role for advertising.

But mass marketing will no longer be the most efficient means of spreading a message. Competitors who learn to target customers—by relevance, not by content or demographics—will increase effectiveness and efficiency and lower their cost. Who has the leading relevance engine? It's not mass-market TV (with its skippable ads). It's not one-size-fits-all, shrinking newspapers. It's not billboards on the road or on web sites. It's Google.

Another reason to still advertise may be to burnish a brand, to help make it cooler because the ad is cool or it appears in a cool place. There is an ongoing debate in media whether brand advertising works online. Advertisers say they do not get the rub-off of branding on the web. They argue that online is a direct-response medium where countable clicks are king and mood can't be conveyed in a banner people ignore. Media people try to convince advertisers that brand advertising does work online— because they charge more for branding and because they don't want to be paid just on clicks. They're both looking at the wrong issue. As *The Cluetrain Manifesto* observed, the internet is filled with human voices of friends and peers, so the artificial, institutional, huckster voice of brand advertising and sloganeering will increasingly be revealed as thin and false. Google's simple, informative, relevant text ads ring truer.

The marketing that is left must evolve. Advertisers are starting to mouth the right words—it's about relationships, not messages, I hear them say. In his 2001 book, *Gonzo Marketing*, Christopher Locke—another coauthor of *Cluetrain*—argued that "the fundamental message of marketing must change from 'we want your money' to 'we share your interests.' In this respect, corporate underwriting is a way—perhaps the only viable way at present—for companies to put their own money where their mouth is." He urged companies to buy ads on relevant blogs—not as a way to distribute messages in banners, but as a way to underwrite blogs, as they would a PBS show. Sponsors say by their support that they share the interests and affections of the blog's readers. Does that co-opt the blogger?

It need not so long as the line between content and ad is clear. Locke also pushed companies to allow employees to blog so they could develop direct, helpful, and human relationships with customers. Robert Scoble, now head of FastCompany.TV, was the poster child for Locke's argument when he blogged from inside Microsoft, in his own voice rather than that of the corporate Borg. He almost single-handedly turned around the reputation of even this company online. Your products and your customers are your ads, and so are your employees.

The best way to burnish a brand is no longer to rub up against media properties like Vogue or the Super Bowl. The best way today is to rub up against people: Sally the blogger or Joe the Facebook friend. The medium is the message and the customer is the medium. Sally is the new Vogue.

Separate the functions of an ad agency today—media buying, research and data, and creative. What happens to each?

Media buying, under Locke's theory, now becomes more important than messaging. When your customer is your ad, media doesn't mean content, it means people. Networks of people will become a force in advertising. Already, media companies, including Forbes and Reuters, are running blog ad networks for marketers. A group of fans on Facebook discussing a product is worth a thousand ads.

Each company must take responsibility for its own research and data. It must know everything it can about its customers and how its products are bought, seen, and used. This knowledge is more than raw numbers derived from snooping on behaviors, commissioning surveys, or quizzing random customers behind focus-group mirrors. We're not data. We're people. So understanding will come from relationships. Ask your customers. Listen. Remember it's a gift economy, and they will be generous if you deserve their generosity.

Creative? Messaging? The more you hand that over to your customers, the better. Apple produces great and entertaining commercials. But in 2004 a teacher named George Masters made a now-legendary commercial for the iPod Mini, filled with psychedelic hearts, that was in some ways more powerful than professional ads because it was made with personal passion.

What becomes of advertising? For the first time, the ad economy may contract. In the past as new media emerged, dollars shifted from old to new—newspapers to TV, TV to internet—but didn't leave the market, according to Bob Garfield, cohost of public radio's *On the Media* and critic for Advertising Age. Garfield observed that while old media shrink, new media are not ready for big advertisers, and big advertisers are not ready for new media. As a result, dollars will disappear into the chasm between. Garfield called this advertising's "chaos scenario."

In addition, as relationships replace advertising, spending will decrease. The new abundance of media online will drive prices down as supply increases and demand decreases. Google's systems will target advertising more efficiently, reducing cost. Opening the market with Google auctions also lowers cost. These savings will not be plowed back into marketing but will need to go toward lowering prices because the internet gives customers unprecedented ability to comparison shop and price will matter more. Some of those savings must be devoted to both improving the product, which now acts as the ad, and improving relationships with customers, who are the new ad agency.

The agency and advertising need to get out of the way in the relationship between companies and customers. Agencies may help solve problems—teaching companies how to build networks with customers, assisting them with product launches—but once the consultation is done, the good consultant leaves town.

Tobaccowala suggested agencies remake themselves as networks. He quoted University of Chicago economist Ronald Coase in his seminal 1937 essay, "The Nature of the Firm"—which is also quoted in *Wikinomics*, *Here Comes Everybody*, and, it would seem, half the business books published lately. Coase reasoned that firms exist and grow when internal friction is less than external friction, when it is easier and cheaper to deal with insiders than with outsiders. "In a networked world, it's easier for us to work with outside people than inside people," Tobaccowala said. "Google, even in its grandiosity, still is a company that believes in forms of partnering." Agencies and other companies, he said, will look more like Hollywood studios, where 80 percent of what goes into a movie comes from outsiders. Google even provides technology to make such collaboration possible. So Google doesn't change just the essence of advertising. It changes the essence of the company. The network is becoming more efficient than the corporation.

Google is an avalanche and it has only just begun to tumble down the mountain. Media was closest to what Google does and so Google's impact on media has been profound and permanent—and it's not over yet. Next in Google's path is advertising. Even though it, too, is close to Google—they are in the same industry—the rumble is only beginning to be heard. Agencies are about to be buried, and they still don't see it coming. The industries we examine next may think they are safe, far away in the valley, under a bright sun. But the Googlanche will hit them, too.

Retail

Google Eats
Google Shops

Google Eats: A business built on openness

What would a restaurant run according to Googlethink look like—other than being decorated in garish primary colors with a neon sign, big balls for seats, and Fruit Loops and M&Ms on every table?

Imagine instead a restaurant—any restaurant—run on openness and data. Say we pick up the menu and see exactly how many people had ordered each dish. Would that influence our choice? It would help us discover the restaurant's true specialties (the reason people come here must be the crab cakes) and perhaps make new discoveries (the 400 people who ordered the Hawaiian pizza last month can't all be wrong . . . can they?).

If a restaurateur were true to Googlethink, she would hunger for more data. Why not survey diners at the end of the meal? That sounds frightening—what if they hate the calamari?—but there's little to fear. If the squid is bad and the chef can hear her customers say so, she'll 86 it off the menu and make something better. Everybody wins. She'll also impress customers with her eagerness to hear their opinions. This beats wandering around the tables, randomly asking how things are (as a diner, I find it awkward and ungracious to complain; it's like carping about Grandmother's cranberry sauce on Thanksgiving). Why not just ask the question and give everyone the means to answer? Your worst diner could be your best friend.

The more layers of data you have, the more you learn, the more useful your advice can be: People who like this also like that. Or here are the popular dishes among runners (a proxy for the health-minded) or people who order expensive wines (a proxy for good taste, perhaps).

If you know about your crowd's taste in wine, why not crowdsource the job of sommelier? Have customers rate and describe every bottle. Show which wines were ordered with which dishes and what made diners happy. If this collection of data were valuable in one restaurant, it would be exponentially more valuable across many. Thinking openly, why not compile and link information from many establishments so diners can learn which wines go best with many kinds of spicy dishes? If you want to be courageous, why not reveal that people who like this restaurant also like that one? Sure, that sends the other guys business—it's linking to them—but in an open pool of information, they will also send business back. Nobody eats at the same place every night (well, there was the time when I went to McDonald's entirely too often). Even a restaurant can think as a member of a network in a linked information economy.

Networks force specialization. In a linked world, you don't want to be all things to all people. You want to stand out for what you do best. That's why chef Gordon Ramsey focuses the menus of the restaurants he fixes on his show, *Kitchen Nightmares*, so they know the business they're in. Serve your niche instead of the mass. Do what you do best.

Now, as Emeril would say, let's kick it up a notch: Open-source the restaurant. Put recipes online and invite the public to make suggestions and even to edit them on a wiki. Maybe they'll suggest more salt. Maybe they'll go to the trouble of cooking the dish at home, trying variations, and reporting back. In the early days of the web, I worked on the launch of Epicurious.com, the online site for Gourmet and Bon Appétit magazines, where I was amazed to see people share their own recipes—there's the gift economy—and also share their comments and variations on the magazines' recipes. For example, a Gourmet adaptation of a bakery's recipe for Mexican chocolate cake brought suggestions to replace the water with espresso (many commenting cooks liked that idea, tried it, and shared their endorsements); double the cinnamon; add Kahlua or rum to the glaze; use cream-cheese frosting instead of the glaze; use neither topping but serve it with whipped cream and berries; toast the nuts; substitute milk and orange juice for buttermilk; coat the cake pan with cocoa powder (helps with the sticking, you see); and even add cayenne pepper

(pepper?). With these adaptations, you could argue the dish is no longer the same; could be better, could be worse. I'm not suggesting that recipes or menus become ballots; see the preGoogle rule about too many chefs spoiling the broth. It's the chef, not the public, who will be held to account if the cake is too peppery. So I'll violate Jarvis' First Law—I won't hand over complete control. But why not gather and use the wisdom of the dining room? A good restaurant has people who appreciate and know good food. It should respect their taste and knowledge, the Google way.

People want to create, remix, share, and make their mark. Perhaps a restaurant could be their platform. Maybe it could stage bakeoffs: Try the chef's version of the cake and Jane's—the winner gets on the menu. The public could suggest dishes they would like the chef to cook: "I had a delicious tart at a café in Vienna and I'd kill to have it again here in Boise." A cook worth her salt would take that as a compliment.

Of course, the best advertisement is a happy customer; this rule is truer with restaurants than with most other businesses. Local restaurants—or national networks of heart-healthy restaurants—can join in relevant conversations and groups online, not to spam them with advertising but to hear ideas and desires and make them come true. Plenty of food fans are already talking online. The FoodBlogBlog counts 2,000 blogs and that's just a start; the U.K. has a Food Bloggers Association; Chowhound.com has outposts all across America. See Chowhound's What's My Craving? forum in New York, in which diners ask fellow diners where to find papusa (thick, stuffed tortillas), a proper Indian biriyani, or Korean jajangmyun (noodles with a black soybean paste). If you think of food as the basis of communities—and it is—then you'll think like Facebook's Mark Zuckerberg and help them organize. Perhaps diners would like to gather parties and you can provide the forum to help. Your restaurant could become the venue for blind dates made on craigslist: get dinner, get drunk, get lucky, get married.

A vibrant online community buzzing around a restaurant will help market it. A social restaurant will soar in search-engine results as diners/users discuss it and link to its recipes. A transparent restaurant that puts much of itself online—recipes, wine reviews, taste data—will also rise in Google search, especially now that Google is making search more local (tell Google where you live and the next time you search for "pizza" it will give you joints in the neighborhood). If people search for where to have a killer soufflé in the area, the name of a restaurant where diners are discussing said soufflé and its recipe should rise as high as the dish.

A Google-driven restaurant won't become a computer-run bistro with the algorithmic menu: roborestaurant. That's not what Googlethink is about. Instead, these tools enable any business to build a new relationship with customers. Not every customer will want a personal relationship; most will eat and run. Following Wikipedia's 1 percent rule, it takes only a small proportion of customers to get involved and contribute great value.

Restaurants are even being crowdsourced. Trend-tracker Springwise reported that a restaurant called Instructables, where customers will make all decisions, is launching in Amsterdam. The Washington Post reported on the creation of an eatery called Elements, whose owners claim it is America's first crowdsourced restaurant. Its volunteers collaborate on concept, design, and logo. The crowd will share 10 percent of the restaurant's profits based on the depth of their involvement. As a fan of sizzling burgers and steaming burritos, I am less than enthralled with Elements' concept: a "sustainable vegetarian/raw foods restaurant" (in the online discussion, there was talk of adding kosher and gluten-free to the mission with round-the-clock breakfast featuring salads and green smoothies). The owner, says The Post, is "creating raw food treats such as oat-hemp balls." I might find a different crowd.

So far, I've suggested that restaurants use the internet to turn the spotlight on diners. Googley restaurateurs can also use the web to become stars. Judging by the popularity of kitchen-based reality shows, I think it's time for chefs to come out from behind the stove. Restaurants have stories, dramas, comedies, and knowledge to share. If I were a chef, I'd blog about my restaurant; my taste, travels, and inspirations; and the trends I see. I'd be blunt and honest. Howard Stern has succeeded on radio and chef Ramsey has succeeded on TV with that formula. So, too, could neighborhood chefs become local stars. I'd make videos teaching people how to cook—remember that the gift economy works both ways. I'd start a cooking club with my most loyal fans—my best customers, my partners—and let them in on discussions if not decisions on the menu and recipes. I might even hand the place over to my community for a night, playing Ramsey in real life and making the restaurant a show. Restaurants don't just sell food—cooked atoms. They are a platform for the enjoyment and discussion of taste. A community and its creativity can grow around that.

Google Shops: A company built on people

Let's visit a retailer who has learned and acted on many of these lessons and is eager to try more. Gary Vaynerchuk, a wine merchant in Springfield, New Jersey, burst onto the internet in 2006 with a daily video blog. Put down this book for a minute—just a minute—go to WineLibrary .TV, and watch one of his shows. Be prepared to be blown back by a jet-engine blast of personality and enthusiasm. Vaynerchuk is hardly the image of a wine snoot. He could just as easily be touting a horse or shouting about his favorite football team (the New York Jets). He's a guy's guy, a man of the people, and that's his point. He's democratizing wine.

Before starting his video blog, Vaynerchuk had already run a successful store with his Russian-immigrant father and family. They rebuilt the place into an impressive, two-story retail space—a library of bottles—and grew revenue from $4 million to $60 million annually over a few years. The Wall Street Journal profiled him in 2006. I'd shopped in his store for years but met him first online.

His video blog made him a star. The show is seen by 80,000 people a day—no small feat for watching a guy holler about wine for 20 minutes and spit his sips into a Jets bucket. His passion is infectious and so his fans spread it around. One day, deep into a show, he mentioned that he was planning an event in his store for his online community. Three-hundred "Vayniacs," as he calls his followers, showed up, flying from California and Florida.

Vaynerchuk got onto big TV thanks to the internet, appearing on *Late Night with Conan O'Brien*, *The Ellen Degeneres Show*, and CNBC's *Mad Money* with its equally forceful host, Jim Cramer. He got speaking engagements. Earlier, I told of his using Twitter to gather a flash party at the South by Southwest conference in Austin. At the conference, he spoke on a panel alongside his Hollywood agent. Then he published a book, *101 Wines Guaranteed to Inspire, Delight, and Bring Thunder to Your World*. The moment it became available to order, Vaynerchuk's fans raised it to 36th place on Amazon's best-seller list. Vaynerchuk started a project to create a collaborative wine—Vayniac Cabernet 2007—concocted with input from his community, who even helped to crush the grapes. (I ordered some. It will arrive after this book is out, so I'll let you know how it is on my blog.) Vaynerchuk understood that he had to provide his community with a platform—a baseball field he called it—so they could play alongside him.

Vaynerchuk told me he wasn't becoming an internet star just to sell wine. He was building something bigger. He was investing in "brand Gary Vaynerchuk." That is why he chose to make his wine shows daily: "Content, baby, indexing in search." Everybody needs *vin de Googlejus*. The more of him there is online, the more he will be found. He is his own ad. The most important factors in retail success used to be location, location, location. Now they are links, Google, Googlejuice. I searched Google for "wine" and Vaynerchuk's store came up on the first page behind only one other retailer, Wine.com, which spent countless millions to build its brand and online positioning. I searched for "wine TV" and Vaynerchuk's show came up first, dominating the listings (where is the Food Network?). In this giant industry, that is nothing short of incredible. He built his brand and market position not with marketing dollars (though his is the only video blog I've ever seen advertised on a highway billboard in New Jersey). He built it with personality, enthusiasm, and relationships in the internet connection machine.

Vaynerchuk is on a mission. "I want to change the way people think about wine and change the way people do business," he told me. On Cramer's *Mad Money*, Vaynerchuk mocked wine and liquor conglomerates for doing nothing socially, acting like monolithic Coke and not like viral brands such as Vitamin Water and Red Bull, which grew by turning customers into advertisers. Vaynerchuk's message: "Social business is the future of our society."

I told Vaynerchuk there were more things I wanted from his store to make it truly Googley. As I shop, I'd love to draw on the wisdom of his enthusiastic crowd and have them recommend wines to me. Wine, as Vaynerchuk says, is always about trying something new. On my latest visit, I came across a Gavi di Gavi. I couldn't recall what Vaynerchuk had said about the variety on his show. I asked a clerk, who told me it was fruity but dry and recommended it. That was helpful. But I don't know this guy and his palate. I'd prefer to have taken out my iPhone and punched in stock numbers to get Vaynerchuk's and his Vayniaks' reviews. Judging their taste by seeing the other wines they like, I would have been in a better position to decide whether to spend that $18. If I'm in a competitor's store that doesn't have a wine the Vayniaks like, I'm now motivated to buy it through Vaynerchuk's growing mail-order business. His customers are his clerks. A store creates value in the knowledge of its customers; that is an unseen asset. It needs to find ways to capture, share, and exploit that value.

After I check out, I'd like a printout of the wines I bought with notes on each so I could choose an appropriate bottle for dinner and share the information with my guests. I'd like a record of my purchases to go online under my account at the wine community Vaynerchuk bought, Cork'd (at Corkd.com) so I could read others' tasting notes and add my own. Vaynerchuk agreed but said that when he first tried giving people cards that tracked their purchases, they assumed they'd be used only to give deeper discounts, not to build content and community. It didn't work then, but might now. Online, I've learned that sometimes an idea doesn't work just because it's tried too soon.

I would love it if the customers could tell Vaynerchuk what to buy. As with the chef in the kitchen, he's still the boss in the cellar. But I'd like to see whether there's a critical mass of Vayniacs who'd say, "Enough with the shiraz already" or "merlot is the new pinot." Perhaps we'd ask him to hunt for a wine: a good Austrian dessert wine for less than $20. He could turn around and ask whether enough of us would be willing to buy that wine to make it worth his effort. Purchasing should become collaborative.

Most of what Vaynerchuk does—or what our dream restaurant would do—could be done in any establishment. Why not expose a store's sales data so I could use that information when I shop? Why not expose my own sales data to me and make suggestions on that basis? Why not gather and share reviews of products so I can make the best selection for my needs and leave happy? Why haven't local stores followed Amazon's lead with these services? In his book *The Numerati*, Stephen Baker says that retailers are only just beginning to think of ways to exploit the data they have about us—like having our shopping carts make personal recommendations.

My wife and I sometimes ask our supermarket to stock a product, but that's a rare encounter with spotty results. Shouldn't the store have forums where customers could ask for products and managers could see when those requests reach critical mass? I know, this suggestion ignores one fundamental economic factor in grocery and other retail businesses—that brands pay fees for shelf space that contribute to stores' bottom lines. But I have to believe that a store that sells me what I want to buy will be better off than a store that sells me what someone pays it to sell.

No local store or chain can compete with the just-in-time, inventory-light efficiency and limitless selection of an internet retailer. So I wonder how the role of the local store changes. Perhaps it becomes more of a show-room run by or for manufacturers. Rather than selling the merchandise right there, it might offer easy ordering and earn a commission. In the chapter on publishing, I looked at printing books on demand. In the chapter on manufacturing, I ask how cars should be sold post-Google. If I were a merchant—a department store, a chain, a local retailer—I'd hope to find a way to curate unique merchandise for my customers as eBay and Etsy.com do for theirs. Maybe a store, like a newspaper, needs to become less a one-for-all clearinghouse of commodity goods and more a pathway to what I really want.

Perhaps a store, like a restaurant, can become a community built around particular needs, tastes, or passions. Look at the data that is created and shared at Netflix and Amazon through sales rankings, automated recommendations, and customers' reviews. Now imagine starting direct conversations among these people. What could be unleashed when Vaynerchuk's customers and fans talk with each other, asking and answering questions, sharing opinions, finding new value in their association with and around him? It's hard to imagine such a community forming around a tire store, of course. But it's not hard to imagine many others where communities could grow: athletic stores (my local store promotes running clubs' events and Nike is holding its own races around the world to encourage such communities to form); food stores (where an instant community of olive-oil fans can gather around choosing which brands to order); electronics stores (if I can read ratings of TVs at Amazon, why can't I see them when I'm in Best Buy?); garden stores (anybody know how to keep the deer at bay?); hardware stores (let's share open-source plans for play-houses); toy stores (any advice for a grandparent buying an eight-year-old boy a video game?); and clothing stores (H&M should have a dating service: "Size 4 petite seeks 34-inch waist, 34-inch inseam, 42-long—no khakis, please").

Community members (aka customers) can become sales agents. Amazon's and BarnesAndNoble.com's affiliate programs enable bloggers to share recommendations. If readers buy, the blogger gets a commission. The online shoe store Zappos has automated recommendation widgets for products. I'd bet Vaynerchuk's community would publish widgets selling their favorite wines. This could become irritating—I don't want my communities to become Tupperware parties. It could be corrupting if bloggers recommend products only to sell them. But the bloggers' brands and reputations are at stake. If I buy a wine you push and it's bad, I won't trust your judgment again. But if I find a new wine I like, I'll give credit to you and to the store that made it possible.

The internet has caused me to go to stores less often. I can't remember my last time in a department store. The mall, where I once browsed, now bores me. Wal-Mart's size scares me. I still enjoy Apple stores but that's often for the education and the free wi-fi and sometimes for the opportunity to ask a fellow cult member for advice. Stores have become dull. Their merchandise is the same and they have less selection than I find online. They are stocking fewer items and running out more often. They charge higher prices than I can find searching the internet. Sales clerks give me less information about products than I can get from Google and fellow customers. And I have to drive to stores, using ever-more-expensive gas and time.

The store's salvation is its customers. Rather than treating the internet as a competitor, retailers should follow Vaynerchuk and use it as a platform. Enable your customers to help you stand out from the crowd. Why should I go to your sneaker store, car dealership, or wine store to buy the exact same merchandise I can find in a thousand stores and sites just like yours? Price will no longer get me there; I can find the best prices by Googling, not driving. Good service? That should be assumed. Information? I'll trust it more if it comes from the community of shoppers. How can you connect with that community? How—to follow Zuckerberg's law—can you help them organize? How—to follow Vaynerchuk's law—can you build a ball field where they want to play? Turn the store inside out and build it around people more than products. Your customers are your brand. Your company is the company it keeps.

Google Power & Light:
What Google would do

Here is our one example of an industry being remade in Google's image that is not hypothetical. Google.org, the company's philanthropic wing—supported with 1 percent of Google's equity and profits—is trying to reinvent the energy industry and with it, our energy economy. It is funding companies and research looking for ways to make power that will cost less than that generated with coal. Their geeky name for the initiative: RE<C (renewable energy cheaper than coal).

Unlike Google.org's other projects—devoted to early warning of health crises, better management of public services, and entrepreneurial growth in the developing world—RE<C is not merely altruistic. It is an exercise in enlightened self-interest. Google and its megaplexes of servers are gigantic consumers of electricity with a growing impact on the economy and the earth. Google is not free of atoms' drag. If Google can help create cleaner, cheaper electricity anywhere it operates, it will improve its own bottom line (the cost of power has been approaching that of the computers themselves in Google's P&L). It will mitigate charges that Google is becoming a major contributor to carbon pollution. Google will have the flexibility to put servers most anywhere on earth, expanding its reach (Google has even patented the idea of wave-powered, water-cooled server farms on platforms in oceans). And the company will get due credit for

helping to save the planet. "Our primary goal is not to fix the world," Larry Page has said. But wouldn't that be a nice fringe benefit?

At the World Economic Forum meeting at Davos in 2008, I attended a forum at which Google's founders presented their energy vision and I came away with a sense of how they would manage other industries and even how they would run the government (more on that later). It gave me a window into the engineers' worldview. Just before this Google.org forum, I had attended a session with Bono and former Vice President Al Gore. They presented their core causes: extreme poverty, debt forgiveness, and disease for Bono; the planet for Gore. The two men tried to insist to the powerful in the great hall that their causes were complementary—can't solve one without addressing the other, they agreed—but in truth, they were competing for the political and economic attention of the governments and corporations there. Gore spoke with passion, even anger, as he insisted that the way to attack global warming is carbon taxes, regulations, prohibitions, sacrifices. He delivered the environmental agenda we've often heard, and did so with authority and determination.

Then I went up the mountain to hear the Google team—founders Page and Brin with Google.org executive director Larry Brilliant. The contrast was stark. To summarize if not oversimplify their vantage points: Where Gore demands taxes and regulation, the Google team proposes invention and investment. Gore and company want to raise the cost of carbon—the cost of polluting—whereas the Google team wants to lower the cost of energy. I'm a bit unfair to Gore, for he would argue that the proceeds of his taxes would fund technology development. But Google doesn't need tax dollars. If it were a country, its $20 billion revenue would rank it about 80th in gross domestic product. It can invest in energy research on its own.

Still, we see different worldviews at work. "You can't succeed just out of conservation because then you won't have economic development," Brilliant said. "Find a way to make electricity—not to cut back on it but to have more of it than you ever dreamed of." More power than you ever dreamed of. Create and manage abundance rather than control scarcity—as ever, that is the Google worldview. Whereas Gore talks about what we shouldn't do, Google talks about what we can do. There we see the contrast between the politician's brain and the engineer's. Google people start with a problem and look for a solution. They identify a need, find an opportunity, and then systemically, logically, and aggressively attack it with innovation.

Page explained that there is a market now for green energy at 10 cents per kilowatt-hour. Some people and companies want to buy it, though it is expensive, because they want to do good or need good PR. But the true market cost of energy is still far below that. Google.org wants to find a way to produce renewable power at three cents per kilowatt-hour, cheaper than coal, which not only gives them a good deal but also shuts down dirty coal plants.

If it succeeds, the foundation would change Google's business and other entire industries, starting with autos. With energy that cheap, Google.org envisions cars plugged into the power grid, solving the problem of pollution from burning gasoline and changing the political balance of oil power (though they point out that the power grid is in woeful need of an upgrade). Google is also supporting an electric-car initiative called RechargeIT, which is trying to accelerate the adoption of plug-in hybrid cars. As a demonstration, Google is converting its own fleet of cars to modified, plug-in Toyota Prius hybrids. Google set up web pages for every car to display data about its energy efficiency—we know how Google loves data. Those cars are plugged into solar-powered charging stations on Google's campus, where the company was producing 1.6 megawatts in solar power by 2008. "It's been great," Brin said. "It produced shade. It reduced cost." Google created a platform for electric-car devotees to make YouTube videos and place them on a Google map, demonstrating popular support and demand for the cars. Google clearly believes it can help create a market for plug-in cars—and why not? It has created new markets for technology and advertising.

Brin said at Davos that Google has an advantage over incumbent oil companies because it does not have a legacy energy business to protect from cannibalization. Still, he was asked, aren't his shareholders going to have a problem with this quixotic investment? The investment is moderate, Page replied, and the payoff is great.

Brin said the foundation's research is concentrating on three energy sources—solar-thermal, deep geothermal, and high-altitude wind—in addition to photovoltaic power. Wind is already as cheap as coal, he explained, but it's intermittent and unreliable on the ground. That's why they're experimenting with high-altitude kites, which operate in constant wind and are cheaper to make than windmills. Deep geothermal requires fundamental research to become viable, but Google.org is making that investment for the long-term.

Though Google has hired experts to work on energy in its own R&D labs, it isn't doing the work alone. As of mid-2008, it had invested $36 million in outside R&D on power in addition to more than $4 million in RechargeIT. Google is not alone in seeing investment opportunities. At Davos, venture capitalist John Doerr of Kleiner Perkins—who invested in Google and sits on its board—held a reception for Bono and Gore (who advises both Google and Kleiner Perkins). Doerr talked about urgent needs and opportunities in energy; by 2008, his firm had raised $1 billion to invest in clean technology.

If Google did run a power company, what would it look like? It would give us all the power we could use at the best price possible, and then it would find ways to take advantage of that. Google could use the power grid itself to distribute the internet and that, too, would help Google, creating more advertising revenue, which could be used to subsidize the cost of our power and access. Google would give us data about our use of power—especially as more appliances become internet-connected. Imagine if every house were to have a web page detailing power usage by every device, as Google has done for its cars. That data would tell us how to conserve (if we even needed to anymore) and it would tell Google how we live (which, in aggregate, will make Google smarter). In his book *Hot, Flat, and Crowded*, Thomas Friedman proposed a similar future with connected devices that manage their own power. If we can generate our own homemade solar, wind, or geothermal power, I have no doubt Google Power & Light would create a marketplace for us to sell power to the grid or donate it to charities. Power could become not only a new market but a new currency.

It's too bad there never will be a Google Power & Light. It's just not what they do for a living. But Google, being Google, may well remake the industry anyway.

GT&T: What Google should do

If only Google ran our cable and phone companies, how much better our lives would be and how much less time we'd spend on hold and at home waiting for the cable guy.

Well, Google has almost had cable and phone companies. The company gives away free wireless internet access in Mountain View, California, its headquarters' town. It has been rumored to be thinking of offering public wi-fi in other cities but denies such plans. It also had been rumored to be working on making its own Google phone. Instead, it created an open mobile operating system, which any phone manufacturer may use (T-Mobile released the first). In an effort to push the Federal Communications Commission and the mobile phone industry toward openness, Google bid in an auction of wireless spectrum in 2008, making a bargain with the government: Google would guarantee a minimum price of $4.6 billion if the FCC required openness—that is, that any device (such as those powered by Google's operating system) could operate on much of the spectrum bought and run by the phone companies. Google didn't win the auction—it won the point. For a few hours, though, it had the highest bid on the table and could have ended up with spectrum and a phone company.

In a forum in Washington, D.C., Larry Page looked a bit dreamy and wistful as he recalled that for a day, his company was in the phone business. Imagine what he could have done with that. At this moment in the show, we should see Page scratching his chin and looking upward as a cloud floats over his head and he ponders an alternative future, a vision based on openness and ubiquitous connectivity. That's the real dream: Google everywhere. Google is constantly nudging to get more internet access for more people at better prices. This campaign is in its self-interest. "If we have 10 percent better connectivity in the U.S.," Page told Reuters, "we get 10 percent more revenue in the U.S., and those are big numbers for us."

Page was in the capital to lobby the government to take the so-called white spaces between TV channels—which become available as the U.S. switches to digital television—and make them freely accessible, like frequencies used for wi-fi. The move would enable the creation of "wi-fi on steroids," which proponents say could give us speeds in the billions of bits a second versus the millions we get now. We could watch, make, and transmit video anywhere. It would goose America's shameful broadband penetration, which in 2007 stood 15th in the world, according to the Organization for Economic Cooperation and Development. U.S. users pay roughly twice what the Japanese do for access that is, on average, 10 times slower, OECD says.

Others don't like Google's idea for the white spaces. The National Association of Broadcasters fought it, saying the plan could interfere with their signals. I'd say they also don't want to make it easy for yet more competitors to grab more of our attention. Cable companies don't like Google's idea; they're making margins as high as 40 percent on internet access and don't want more competition and disruption. Phone companies don't like it as they're just getting into the cable business. Mobile phone companies don't like it, for once we get broadband on any device, we can use it to do anything, even make phone calls from the web without paying for minutes. With open devices—the ones Google insisted on with the FCC and the ones Google is enabling with its mobile operating system—on open networks we can kiss our two-year contracts and early-cancellation fees good-bye.

Telecommunications is a perfect arena for Google because it's ripe for disruption in business models enabled by technology—Google's specialty. Google doesn't want to be in the wires-and-pipes business, but if our connectivity were freed from its constraints, Google would benefit. We'd spend more time online. We'd create and consume more. Google would have more to search and organize. Google would serve more ads. It would make more money. We would spend less money. It's a magnificent conspiracy of Google and everyone who opposes telecommunications oligopolies.

Who wouldn't like to stick it to the cable guy? The American Customer Satisfaction Index from the University of Michigan said in 2007 that cable and satellite TV suffered "the lowest level of customer satisfaction among all industries covered." The survey attributes some of the problem to the monopolies these companies held and the pricing control that allowed: "Comcast is one of the lowest scoring companies in ACSI. As its customer satisfaction eroded by 7 percent over the past year, revenue increased by 12 percent. Net income went up by 175 percent and Comcast's stock price climbed nearly 50 percent." Let's replay those numbers: Even as we hated the cable company more, its revenue, profit, and stock price all climbed. That might work today. But Wall Street must someday learn that angering customers is not a sustainable business model.

Advertising Age's Bob Garfield got angry at Comcast over his simple effort to get service at home. Garfield—who has confessed to envy of my Dell hell—launched a crusade against Comcast in a screed in Ad Age, in a podcast, and on a blog called Comcast Must Die, where he urged customers to share their nightmares. "Congratulations," he told them. "You are no longer just an angry, mistreated customer. Nor, I hope, are you just part of an e-mob. But you are a revolutionary, wresting control from the oligarchs, and claiming it for the consumer. Your power is enormous. Use it wisely." Comcast responded by assigning a vice president to read blogs and Twitter and deal with complaints and problems there. That helps, but it doesn't solve the essential problem: Cable companies exist to frustrate us. I responded to Garfield on my blog, suggesting that a more constructive approach might be to help Comcast reinvent itself.

What would an ideal cable and telecom company—Google Telephone & Telegraph—look like? First and foremost, it would be a platform that exists to help us do what we want to do. More than making calls and consuming content, it would turn the pipes around and help us create, share, and sell. It would be the home and host of our ambitions. Just as Google bought Blogger, allowing us to publish, and YouTube, allowing us to broadcast, its cable company would be our personal technology platform with tools to create content, products, and even companies. If we succeed, it succeeds.

Even if we did not have such creative ambitions, Google would still provide no end of services in our personal computing clouds. It does that already with Gmail—the best webmail and best spam-fighter out there; Google Docs—free and collaborative word-processing and spreadsheets; Google Calendar; Google Maps; Google Apps. If Google were my local cable and phone company, I'd expect it to provide the means for me and my neighbors to join groups and share information (which is what local newspapers should be doing as well). My neighborhood and town should be searchable. Google has started playing in the local arena with maps, news, and ads, but imagine when Google becomes truly local and I get a version of it and its services tailored to my area, my office, even my family and house.

GT&T would be open. Gone soon will be the days when a company can make its money by telling customers what they cannot do, as cable companies long have done (you can't upload that much; you can't watch the shows you've already bought without paying extra for our on-demand service; you can't put as many TVs in your home as you want without paying more; you can't watch a TV without adding our cable box; you can't buy just the channels you want but instead have to buy overpriced bundles; you can't get a time when our technician will actually show up . . .). Google knows that the more we use the internet, the better off it—and its hypothetical cable and phone company—would be.

I think GT&T would give us portability: Just as I can get my email anywhere on any device thanks to Gmail, I should be able to get the video programming I paid for in any room of the house or even in a hotel room across the country—without the need for cable boxes, TiVos, or Slingboxes. My inability to do this today isn't fully the fault of the cable company; it's the result of archaic notions of copyright protection designed for outmoded technology. Studios and networks have argued that it would be a copyright violation if the cable company kept a copy of a movie I'd bought on its server so I could get it from anywhere. But there's hope this practice will change after an appeals court decided in 2008 that remote storage is not a violation. The other barrier to portability is hardware. Cable companies are in the business of renting cable boxes to us, thus amortizing their cost and giving them control. Cable companies don't see how renting boxes limits them, adding to their capital outlay, delaying technical improvements, and reducing our use of their service. GT&T would put forward open standards for everyone who makes TVs or video recorders, eliminating boxes and enabling anything to plug into the network, as on the internet. Google followed that model of openness—increasing use and utility—when it released its open-source browser, Chrome. There is hope on the hardware front as consumer-electronics and cable companies have at long last agreed to allow some integration of devices.

Google would understand that in a larger network of content and information, its opportunity would be to help us find what we want. It would provide a guide to cable as it provides a guide to the world's information. GT&T would become the new TV Guide and the new TiVo mixed in with a search engine and a social network. Where would it get that guide information? Where Google gets it now: from us, from the crowd. We'd all be networks, recommending shows to each other, no longer caged by the taste and schedules of a few networks. We'd act as a mass of niches, not a mass. No doubt Google would analyze data about our actions and taste and feed that back to us as recommendations, as it does today in search. Why wouldn't GT&T become the great personalized search engine of entertainment, the Google of culture? If somebody else doesn't do it first, they probably will.

One doesn't think of Google as a customer-service company. Its stuff just works. I rarely hear people complain about them as we do about phone and cable companies. After I told Doc Searls, another coauthor of *The Cluetrain Manifesto*, about my book, he blogged about his customer-service experience with Google. He needed to register a domain and if you've ever done that—at Network Solutions, GoDaddy, or other sites—you know that it can be a cluttered maze of attempts to get you to forget to click on boxes so you get charged for extra services. (It's a variation on an old sales trick: When I worked at Ponderosa Steak House as a teen, we were taught to raise a ladle of canned mushroom gravy over a diner's steak and ask, as if "no" were not an option, "Mushroom sauce?") "Without exception, my experience with domain name registrars has been an upstream slog against a torrent of promotional distractions," Searls wrote. "Nobody hates white space more than a domain-name registrar." But when he discovered that Google offered this service for $10, he used it and in minutes, was done. "I used Google because I trust them not to treat me like cattle—or worse, as a potential sucker. . . . I bought this domain name from Google because I have a mutually respectful relationship with them. That relationship does not require human involvement, but it does require human values. Especially respect."

GT&T would make a compact with customers to provide reliable service. When it fails, we could use Google's own tools against it. We could put up a Google map that we all fill in when we have trouble with our cable. We could record our conversations with customer-service people and put that and our complaints on YouTube, searchable via Google. We could share how fast our bandwidth is at every address and publish it all in a Google Docs spreadsheet. Google would know that it couldn't fight us or win trying. Google is a platform for watching Google.

Would we ever have to wait all day for the Google cable guy to show up? No. If "cable" were wireless and worked with any device that met open standards, there'd be nothing to string to our homes, nothing to install, nothing to come fix. We could choose to use our bandwidth just as we wanted, as we use our power and water at will. I want a cable company that follows Jarvis' First Law. Wouldn't that be novel: control in customers' hands?

How would GT&T profit? How else? Advertising. It might still have to charge us for bandwidth and services. But Google would be smart enough to create new means to target local and national ads to us, using that revenue to subsidize the service so it would cost us less and we would use it more. Thus GT&T would make yet more money: a virtuous economic circle. Bandwidth could be free if what we do with it has enough value.

I wish Google would change its mind and get into the cable and phone business. But if it doesn't, there's no reason our cable torturers should not operate as I've outlined. You don't need to be Google to act like Google.

The Googlemobile: From secrecy to sharing

I sat with carmakers some time ago and suggested what I feared was blasphemy: I urged them to open up their design process and make it both transparent and collaborative. Car companies have no good way to listen to customers' ideas. If they had, years before, I would have been among the legions who'd have gladly told them they should invest 39 cents in a plug for car radios so we could connect our iPods. Every time I try to listen to podcasts in the car via various kludges—FM transmitters that couldn't transmit an inch away and cassette-tape gizmos (if you still have a cassette deck) that are loud and unreliable—I curse car companies and their suppliers. At least let us help design the radios you install, I urged.

My plea was sacrilegious because automakers have long been secretive about design. Design and surprise, they think, are their special sauce. That's why they cloak new models like classified weapons, setting off games of cat-and-car with photographers who try to scoop the secrets. Apart from the most fanatical car fan, do the rest of us still care? The anticipation I remember about a new year's cars—like a new season's TV shows—is gone. Cars have lost their season. They stay the same year upon year. They all start to look alike. They rarely engender excitement. How could a car company reinject affection into its products and brands—how could it get a little love? By involving customers, I argue—by turning out cars customers want because they had a chance to say what they want.

Internet analyst Jeremiah Owyang compiled a list of auto industry social-media efforts on his blog: Some automakers let customers make their own ads for cars, make their own emblems, or color pictures of cars. GM's vice chairman, Bob Lutz, blogs. Chrysler has solicited customers' ideas—but in a closed form that prevents them from commenting on each others' suggestions. Chrysler also created a customer advisory board of 5,000 selected drivers. The Mini has its active community of owners.

The problem with these efforts is that they do not allow customers to openly affect the product. Perhaps one of the ideas presented to Chrysler in emails or discussed in the Mini's community might influence a decision that will come off the line in a few years. But we'd never know it. Indeed, the companies' efforts at interactivity work hard to keep the customer from doing harm. This is interactivity as defined by a children's museum: Here are the buttons you may push without breaking anything; knock yourself out, kids. But just as companies should hand over their brands to customers so should they hand over their products.

What if just one model from one brand were opened up to collaborative design? Once more, I don't suggest that design should be a democracy. But shouldn't design at least be a conversation? Designers can put their ideas on the web. Customers can make suggestions and discuss them. Designers can take the best ideas and adapt them, giving credit where it is due. I don't imagine customers would collaborate on transmission or fuel-pump design—though a few might have great suggestions if given a chance. But they would have a lot to contribute on the passenger compartment, the look of the car, the features, and the options. They could even get involved in economic decisions: Would you be willing to give up power windows if it got you a less-expensive car or a nicer radio? This collaboration would invest customers in the product. It would build excitement. It would get the product talked about on the web and linked to and that would earn it Googlejuice. It could change the relationship of customers to the brand and that would change the brand itself. Imagine that: the collaborative community car—our car.

A car company could take any existing brand and model and work with the community that already exists around it. Go to Facebook and you'll find communities of greater or lesser involvement and affection around many car brands. I lost count of the Facebook groups for BMW when I hit 500. They included, with more than 800 members, the "If the BMW M5 was a woman I would marry it" group in addition to the "I hate BMW drivers, they are all c—ts" group with 510 members and, with

446 joiners, the "I piss people off, cause I drive a BMW" group (don't invite the latter two to the same party). At Meetup, there are six clubs where people gather with their Beemers. BMW has its own official car club offering 75,000 members rebates on cars and discounts on Brooks Brothers clothes (do they see the demographic humor in that?). These are the company's best customers, its partners. BMW should solicit their help in designing cars, supporting fellow drivers (there's a little of that in the club forums), and even selling cars.

On Facebook, BMW invited customers to color pictures of its car. It's hard to imagine something more children's museum-like than a company enticing adults to color cars. But more than 9,000 people submitted their designs in only a few days. What that tells me is not just that they love their BMWs but that they would love BMWs that looked unique—BMWs that expressed their muses as well as their libidos. What an opportunity the industry has to bring humanity and personality back to cars. If so many of us like to express ourselves in blogs; YouTube videos; Facebook, Bebo, and MySpace pages; and Flickr photos—if, as Google understands, many of us want to have a strong identity online through self-expression— why wouldn't we want to express ourselves through our cars? Companies have turned their products into commodities by imposing such sameness on them. I know, it's about efficiency—four lines of cars built under four brands on the same body with the same engine and parts makes them cost-effective. Factory efficiency and dealer economics also stop us from ordering custom-made cars anymore. We buy off the lot, not out of the factory, and we buy cars that are often loaded—like cable subscriptions— with things we don't want. (Every time I start my car, I turn off the night-vision rearview mirror, a $100-plus option I didn't want but had to buy.) Sure, there's an aftermarket for options—piney scent strips, hubcaps that spin, mud flaps with mirrors in the shape of naked women—but, well, that's just not me.

Toyota's Scion took a small step toward personalization when it enabled drivers to design crests for their cars. Now go the next step and imagine I could take an unpainted car to any of those BMW designers on Facebook or my student the graffiti artist and have my car painted so that it looks like no other. It'll cost me. But I'll bond with that car and love it because it's an expression of me.

That unpainted car would be the beginning of an auto company thinking open-source. What if the company also produced a car onto which I could graft someone else's dashboard or seats or grill or engine? Earlier, I talked about Google replacing its fleet of company cars with Toyota Prius hybrids that were modified so their extra batteries could be recharged with solar power. That is the Googlemobile. Google treated the Prius as a platform. Toyota should be delighted. It should build in opportunities to modify its car in countless ways. I can hear the objections: It could complicate production, raise costs, and confuse brands. Maybe. But it could give me the car I want. The car company of the future should be a platform for more car companies that make the automobiles drivers want, not the ones they settle for.

There are projects aimed at building the open-source car, among them Oscar from Germany, the c,mm,n (or common) hydrogen car from universities in the Netherlands, and the Society for Sustainable Mobility car (being built with 150 part-time engineers, according to Fast Company). The Aptera from Bill Gross' IdeaLab (more from him later, in the chapter, "Google Capital") is a beautiful, three-wheeled hybrid or electric vehicle set to launch in California. Tesla Motors is building a six-figure-plus all-electric sports car with funding from one of PayPal's founders. They are all cool and I wish them luck. But it's damned difficult to get a car company operating at scale—ask John DeLorean.

That is why I think a car company that already operates at scale should think open-source and welcome these nascent efforts to build atop them. Imagine seeing a million Priuses, Saturns, Fords, or Apteras on the road and wondering what's inside each, what makes it run, who painted it, where you can get that great grill. Imagine being given the power to customize your car from the ground up. Cars would be exciting again. Give me control of my car and I will own that brand, make that brand, love that brand, and sell that brand because it is mine, not yours. That will be the key to marketing Googlemobiles: passion, individuality, creation, choice, excitement, newness. Drivers will start Facebook groups, blogs, and Meetup clubs extolling the wonders of the cars they choose—no, make. Outside product designers and manufacturers will accessorize and improve the open-source car—as outside developers make Facebook apps and mash up Google Maps—which will support new businesses and help sell more cars. There is the advantage of being a platform.

Now we come to the big problem facing car companies: dealers. We don't like car salesmen (in a 2007 Gallup survey, Americans rated them at the bottom—tied with lobbyists—with only 5 percent saying dealers had high honesty and ethics). They add little or no value to the transaction and none to the product. They make buying a car uncomfortable. Car companies in the U.S. are stuck with franchise laws that won't allow them to sell directly. So what should they do? I suggest they start by creating a platform for customers to say just what they think of car salesmen so companies can rub dealers' noses in it. Perhaps the voice of the people will reach and convince Congress to deregulate and open up car sales. We now do most of our car shopping online. We comparison shop, read reviews, review specs, and talk with friends. All we need the dealer for is a test drive. Once I know I want a car, why should I have to drive to the dealer; why doesn't the dealer or a manufacturer's representative deliver the car to me? Why can't I buy a car at the auto show? Why should I have to negotiate with three dealers for the exact same product when open pricing online has already told me what the market will bear? The dealer structure builds in inefficiencies and costs that the industry—and we—cannot afford.

The repair system is little better. My warranty is really an insurance policy that I should be able to redeem at any repair shop. The car company could still provide training—I would prefer to go to someone certified—and would sell parts. If the repair market were more competitive, the car company and I would each benefit.

I discussed my rationale for the open-source car platform with Fred Wilson, a venture capitalist you'll hear from shortly, and asked him what a Googley car company would look like. He said it already exists. It's Zipcar, which provides 5,000 cars to 200,000 drivers in various cities and campuses. Drivers join Zipcar for $50 a month, then make reservations online and pick up a car in any of a number of garages, paying $9 an hour or $65 a day in New York, including insurance, gas, and 180 miles. I can get similar rates from traditional rental companies but with less flexibility and convenience. Zipcar says each of its cars replaces 15 privately owned cars and that 40 percent of its members decide to give up owning a car. Similarly, Paris' mayor announced in 2008 that the city would follow its successful bike-sharing program by making 4,000 electric cars available to residents to pick up and drop off at 700 locations. The goal is to get Parisians to buy fewer cars.

I know what you're thinking (and can hear the peals of laughter all the way from Detroit): The last thing a car company should want is fewer cars. Are you nuts, Jarvis? Are you a communist or some tree-hugging fanatic? No. I'm just turning the industry upside-down. When I put the question to adman Rishad Tobaccowala, whose agency works in the auto industry, he said Detroit is not really in the business of making cars. He channeled the Googley car company and said: "I'm in the business of moving people from place A to place B. How can I do it in different ways? And as they are moving from place A to place B, how do I make them feel secure and connected?" He said that except for sleep, we spend more time moving around than at home. "Screw Starbucks as the 'third place.' The third place today is the automobile." What is the automobile really about? "Navigation and entertainment," he said—not necessarily manufacturing. Indeed, Tobaccowala said the most interesting parts of the General Motors business had been OnStar and—credit crunch aside—financing. Manufacturing is expensive, vulnerable to commodity pricing, labor-intensive, weighed down by gigantic benefit costs, and competitive. There's the tyranny of atoms.

What if a car company became the leader in getting people around and used others' hardware: planes, trains, and automobiles? You tell the system where you need to go—or with access to your Google Calendar, it just knows—and it gives you choices at various price points: Today, you can take the train for less. Tomorrow, you drive because you're running errands. The day after, you carpool to save money. This weekend, you get a nice Mercedes for the anniversary dinner. Next week, you take a chauffeur-driven car to impress clients. Along the way, you can pay for options: your entertainment synced in the car, wireless connectivity on the train, alerts to your iPhone, navigation concierges who direct you around jams. This is the new personal transportation and connections company built on the old car company as a platform. Hop aboard the Googlemobile.

Google Cola: We're more than consumers

If big cars are hard to Googlify, packaged consumer products are harder. They are the building blocks of the mass market, predicated on manufacturing efficiency and marketing to a critical mass. Since the beginning of the internet as an advertising medium, it has been a truism that no one will click on a banner ad for—let alone join a club or write a blog post about—toilet paper. TP is everyone's example of a product that could not possibly benefit from the web. There's nothing Googley about toilet paper. Right? Besides perhaps getting TP printed with Wikipedia's knowledge (there are TP publishers) or made from renewable, recycled resources, I must concede: I can't conceive of Google Ultra Soft Toilet Tissue.

Are all consumer products doomed to life without Googlification? Let's imagine Google Cola. The strength and weakness of cola, like other consumer products, is that it is intended to be one-size-fits-all. Yes, a number of cola brands and variations fight for scarce supermarket shelf space. But there are never enough varieties. I can't find my perfect cola. Mine would be caffeine-free but made with sugar instead of artificial sweeteners (can't stand the aftertaste) and it would come in a small can so it wouldn't go flat, or better yet, a bottle that could be reused. It might have a flavor added (cherry today, coffee tomorrow). I'll take Coke or Pepsi (I'm bicola), but I don't like off-brands (I still shudder remembering Howard Johnson's HoJo Cola). What if Coke retooled a bottler to make special-order batches to be delivered—but only if I committed to buying so many cases a year? I would pay a premium to subscribe to my perfect cola.

If I sold this Jeff Cola to others on my blog or in the neighborhood (convincing them that decaf coffee-flavored soda is not an oxymoron) perhaps my price could drop because I'd be bringing in more sales and volume. I'd create a cola club. It's no different from Gary Vaynerchuk's Vayniaks making and promoting their own wine. We'd become product managers and salespeople as well as consumers and customers. We could invent our own flavors of Coke, sold under our brand, using Coke as a platform for manufacturing and distribution. We'd be in the cola business. Will my cola go mass? Not a chance. But a bunch of smalls could add up to a big, and Coke ends up in a new and loyal relationship with a lot of customers. It learns more about the public's taste and may develop new products to sell on a larger scale. It saves on marketing as collaborators sell products. It gets a piece of businesses that might otherwise take bits of market share. It finds a way to battle the tide of commodification in consumer products and joins in the small-is-the-new-big economy.

The cola strategy could be applied to most any consumable product that would benefit from specialization and personalization: cookies, candy, ecological home-cleaning products with personalized scents. It could be executed not just by big companies but also and more likely by small ones using sales platforms such as Amazon and eBay. About the only mass product I know that customizes today is M&Ms, which you can order printed with a photo ($39 for 21 ounces) or a custom color ($48 for 56 ounces). That's a nice gimmick, but it doesn't change the essence of the product. What if I could get coffee-flavored M&Ms or my decaf coffee-and-M&M-flavored soft drink in bottles for me and the hundred people I found like me? That would be Google Cola.

How about gadgets, then? Personal electronics might seem immune from Googlification because they are so complex in engineering and manufacturing. Yet technology is also what makes gadgets easier to change than cars, as a device can be updated via software instead of hardware. That's what Google is doing by offering its mobile operating system to any phone maker.

I could see Google proposing open standards for no end of connected devices. We can already buy refrigerators with internet screens. Their fabled promise is that someday they will take inventory of what's inside, telling us what we can make with what we have and automatically ordering what we need. That's the kind of information Google would love to organize. Home-delivery services Fresh Direct and Peapod in the U.S. and Tesco in the U.K. could order and deliver what we need and give us coupons for related products. Epicurious.com could suggest recipes based on what's in the fridge. Refrigerators become platforms for these companies to serve us.

We have connected home-security systems with sensors and cameras. We have connected home-entertainment systems that can pipe web radio stations, iTunes music and movies, and YouTube videos to any device in the home. We will have connected cars with links to traffic information and feeds of entertainment. We have cameras connected to GPS satellites and to our computers. We have mobile phones that are becoming computers. Any device that produces information, that can be personalized or adjusted, or that communicates with or entertains us will be connected to the internet and to Google. Google will listen to and speak through these gadgets—if we give it permission—and deliver related information. Google would love to use that information to give us highly targeted and relevant advertising. That might freak privacy warriors. But if we can control that flow and benefit from it (with relevant content and ads, bargains, and subsidies for the services we use), I'd hook up my fridge and phone. Google could become the operating system not just for the web and the world but for our homes and lives.

Another challenge: fashion. We know what Googley fashion is: T-shirts, shorts, and sandals. It's hard to imagine spartan, garish, geeky Google having an impact on taste and trendsetting, which are decreed by designers, fashion editors, and Hollywood. Fashion is top-down—or it was. Just as the internet democratizes news and entertainment, it is opening up style. A darling of the open fashion movement is Threadless, a T-shirt company that invites users to submit designs, which are voted on, Digg-like, by the community. Winning designers receive $2,000 plus a $500 credit and $500 every time a design is reprinted. They become the Versaces of the crowdsourced runway.

Just as in entertainment, we are learning that the public wants to create and leave its mark. A smart response is to create a platform to make that possible. CafePress.com and Zazzle provide the means for anyone to make and sell designs on T-shirts, mugs, bumper stickers, even underwear, getting a cut of every on-demand order. Threadbanger, a weekly internet video show, teaches viewers how to make cool do-it-yourself fashion with young designers. See also BurdaStyle.com's open-source sewing from the German publishing empire Burda, which decided to take copyrights off its dress patterns and invite the public to use them, adapt them, create their own, and share them. The site is filled with patterns, how-to's, and discussion. Springwise reported that SANS, a small New York label, stopped selling its hit $85 square shirt in 2008 and then released the pattern. For $6, you get the pattern, which you print out at home, and a SANS label to sew inside. Opening the design is a nice idea but I can't sew. So craftsmen could build a business out of making SANS or Burda designs on order, as some are doing, selling them on Etsy, a site filled with unique, handmade items, which has been the store for more than 100,000 sellers since 2005.

OK, consumable goods, gadgets, and fashion could be Googlified. But what about Google TP? Surely it is not possible to bring Googlethink to toilet paper. There won't be communities around toilet paper. I shudder to imagine TP 2.0 after seeing a commercial for toilet paper whose USP (unique selling proposition) is that it doesn't leave little paper bits on your butt. Boy, that must have been a tough sales conference. I can't think of a better reason for advertising not to exist.

As with newspapers, perhaps it's time for the TP industry to get out of the paper business and ask what business it is really in. Cleanliness, right? When I was in Davos, what amazed me almost as much as hanging out with heads of state and industry was seeing an automated, self-cleaning toilet seat in the conference center. After flushing, a motorized arm comes out and grabs the seat, cleaning it as it rotates. It's mesmerizing. I took video of it to share on YouTube. (Google "Davos toilet" for my video. Or for a more entertaining if politically incorrect demonstration, search on YouTube for "Swedish toilet seat Gizmodo"). The company that makes that product is not in the paper business. It's in the clean-seat business.

Toto, a Japanese plumbing manufacturer, has decided that the business is neither paper nor clean seats but clean bums and happiness. Toto invented the Washlet automated, computerized toilet seat, a marvel of technology that heats the seat to a cozy 110 degrees and spritzes you with warm, clean water after you've done your business. Then it dries you with gentle, warm air even as it magically eliminates odors. (On YouTube, search for "Toto Washlet FlushTV" to see a demonstration by W. Hodding Carter IV, son of the former Carter administration official and author of *Flushed: How the Plumber Saved Civilization*.) Before you laugh, know that Toto has sold 17 million Washlets (they advertised on my Buzzmachine with smiley faces superimposed on naked, happy, clean butts). The Toto is hot on YouTube with videos that have tens, even hundreds of thousands of views. Hollywood actor Will Smith has bragged on TV that he has the deluxe, $5,000 model and doesn't spend a dollar on TP. Here we have the perfect convergence of problem and solution, hardware and software, technology and life with bottom-up marketing. This is the post-TP Googley toilet.

Even in atom-based enterprises, the connections the internet makes possible can bring business benefits. No end of consumer products would be helped from a more open conversation: tool makers listening to craftsmen, cooking-utensil companies opening up to cooks, athletic equipment companies watching out for what athletes and trainers want. One should find opportunities to make more targeted products and to partner with customers to design, support, and sell products. Google and the internet change everything, even factories.

Google Air: A social marketplace of customers

In contemplating how to remake an airline with Googlethink, I had just about given up. What can one do with such a commodity service, particularly one that has deteriorated so badly? Air travel's business model today is based on overselling seats, billing us for checking bags, charging for pillows and pretzels and just about everything they can think of but air, jamming planes to the point of torture, treating customers as prisoners who can be kept on runways for hours without the food and water an inmate is allowed, and withholding information—all the while raising prices. Google couldn't fix that. No one could.

But then I applied Google rules about connections and the wisdom of crowds with Zuckerberg's law of elegant organization and my own first law and asked how travelers on planes, trains, and ships or in hotels and resorts could be given more control (of anything but the cockpit, of course). And I wondered, what if passengers on a plane were networked? What if a flight became a social experience with its own economy?

Start here: Most of us are connected to the internet on the ground. Soon, we'll be connected in the air as planes, like hotels, finally get wireless access (after earlier failed attempts). Wi-fi is good for airlines because they will have something new to charge us for and because it will keep passengers busy and perhaps less likely to grumble and revolt at delays (though we might just blog and Twitter every problem and indignity as it occurs). Once connected with the internet, passengers could connect with each other. It would be easy for the airlines—or passengers themselves—to set up chats and social networks around flights and destinations so we could hook up before and during a flight. We could organize to share cab rides once we land, saving each other money. We could ask fellow passengers for tips about restaurants, museums, and stores and ways to get around. If the wi-fi were reasonably priced and if there were electric plugs at our seats, we could also spend hours happily playing games with each other.

Back when the 747 was introduced, it was supposed to offer lounges where passengers could hang out together. That didn't last long as every inch was soon crammed with revenue-producing seats. Lounges are supposedly set for a comeback in the new Boeing 787 Dreamliner and Airbus A380 superjumbo jets. So imagine if in our onboard, online social network, we could find people we want to meet—colleagues going to the same conference, travelers with shared interests, future husbands and wives—and we could rendezvous in the lounge. The flight becomes a social experience.

I know this vision sounds far-fetched given our current experience of air travel. But play along. Socialization could be a key to decommodifying the airline. What if passengers chose to fly on one airline vs. another because they knew and liked the people better? BMW drivers mingle with each other on Facebook; Lufthansa passengers could do likewise and they'd have more in common—shared affection for travel and for a destination. Remember: Your company is the company you keep. Your customers are your brand. Airlines might want to encourage more interesting people to fly with them because interesting passengers would attract interesting passengers. Airlines could offer discounts and benefits to people who are active and popular in the social network. Today, airlines offer only seats: commodities. What if, instead, they were to offer experiences and societies? I know, the last thing we want most of the time is to get stuck with a talkative twit in the next seat. Maybe that's because, by the time we get on a plane, we're in rotten moods. Suspend disbelief still. Imagine returning to the days when we met interesting people in chance encounters in the air. Maybe passengers could choose to sit next to each other. Next to the right talker, I might tolerate a middle seat. It would probably have to be David Letterman or Oprah sitting next to me. But it could happen.

These passenger networks raise the possibility of creating a new economy around the flight. Airlines could set up auction marketplaces for at least some seats, as JetBlue began doing experimentally on eBay in 2008: What's it worth for you to fly to Orlando next Monday? Rather than buying seats only from the airline, if late-booking passengers could also buy seats from fellow customers in an open marketplace, that could solve some of the airlines' overbooking problems, reducing the need to pay bumped fliers. Yes, speculators could arbitrage seats, but if they're paid-for and nonrefundable, what problem is that for the airline? Resellers become market makers. This exchange sets a new market value for seats that in some cases will be higher than the airlines' own fares.

The airline could use the exchange as a prediction market to forecast and maximize load. It might see a surge in demand for a destination, perhaps for reasons it could not predict (a new conference or festival, good media coverage for a getaway, a travel bargain, or currency fluctuations unleashing pent-up demand). With sufficient notice, the airline could add capacity, which would keep it ahead of arbitrageurs. The airline always controls supply and now it would know more about demand. Similarly, if a flight were light the airline could offer passengers alternatives at big discounts to enable it to cancel the flight and reroute equipment long before departure, creating savings at the bottom line. The airline would increase efficiency and profitability; the passengers would get a dividend; and the environment would get a break. An open and flexible social marketplace could transform the airline economy.

Why shouldn't airlines also turn frequent-flier miles into an open market? In these miles, airlines created a virtual currency with greater reach and value than the fake currencies of Second Life or Facebook. But miles are essentially illiquid. Airlines make it next to impossible to get frequent-flier seats unless you're flying to Krakow on Christmas Day a decade hence. Their other offers—use miles to buy a TV—are bad deals; Google search tells me so. Miles have been devalued to the point that they offer ever-lessening incentives to choose one carrier over another; they no longer act as the decommodifier the airlines intended. So open it up: Let us bid on frequent-flier seats, upgrades, and silver status with miles. Let us barter miles with each other (I'll sell you this iPod for miles I need for my vacation). The currency would regain value. It also means more miles will be redeemed, but that sword hangs over airlines' heads in any case.

These exchanges bring subtleties. In some cases, I won't want to reveal my identity (telling strangers I'm leaving town); in others, I will (because I'm doing business). As seats are traded, identities and credit cards must be in the system for security. And so on. Yet creating such a network could rebrand the pioneer as the social airline—the fun airline, the nice airline, the airline where I'm back in control. (I would use the social network to start a movement to save my knees from the asses who slam their seats back into them. In an open market, I might even pay them not to.)

Now imagine if airlines used these networks to capture the knowledge of their smart-about-traveling crowds and convert that wisdom into value. On our return trips, airlines should ask us to rate and review the hotels and restaurants we frequented. They should ask natives to share insider advice on eating and shopping in their towns. Similarly, hotels should capture guests' reviews of nearby restaurants (as Hyatt has begun doing with its Yatt'it travel community), and cruise lines should gather shopping tips for every port. Travel companies have currencies to pay for the information: They could reward us with frequent-flier miles or discounts on our next trips. And because they know who we are, they could anonymously aggregate data to enhance the information, as I suggested for restaurants: "American Express Platinum customers recommend . . . ." Or: "Canadians traveling to Florida really like . . . ." Airlines could collect an incredible database of live knowledge from real travelers with fresh information. Over time, they'd outdo TripAdvisor and Fodor's—or the airlines could supply them with branded content, which in turn promotes the airlines. The airlines themselves become publishers by listening to, gathering, and sharing the knowledge of customers.

The key to remaking an airline in this mold is giving control, respect, and organization to the customers, helping them find each other and organize into conversations and markets. The customers have value to give. Airlines can capture that value in new ways to improve prices and the bottom line (see the discussion of Ryanair in the chapter, "Free is a business model"). But passengers won't give their value if they are not valued, if they are still treated as cattle and criminals.

At a party at the World Economic Forum at Davos, when I met one of the Google cofounders, I mentioned that I was exploring the idea of what an airline would be like if Google ran it. I said I thought it would be social. He grinned and told me about a technology entrepreneur who had founded just such a social airline, but it had to shut down when employees were caught in a scandal smuggling drugs. Pity.

Google Real Estate: Information is power

I've already aired my enmity for real-estate agents and their oligopolistic fee structure. So I start this chapter not by suggesting how they can remake themselves but instead by speculating how others can disrupt, undercut, and destroy their business.

I should explain why I feel this way about agents. I have had some bad eggs. I also know there are nice agents. It's not personal. It's financial. I do not believe that agents add 6 percent's worth of value to a home sale. The only reason they could demand that commission is because they have controlled the multiple-listing service (MLS) that is key to having your house seen by buyers. Agents aren't the only parties ripping us off in the process. Title insurance is particularly irksome, as is the necessity of having surveys done and redone, as are home-inspection rackets that have never found the flaws I have found after moving in. Let's not forget lawyers, who make the process unnecessarily complicated so they, too, may soak us. Then there are newspapers that charge too much for inefficient advertising.

The real-estate business is ripe for disruption. Attempts so far have failed because they only try to break open the existing structure, to create discount brokerages that can get homes into that precious multiple-listing service. Even though the U.S. Justice Department in 2008 reached an antitrust settlement opening up the MLS to discount brokers, we are still trapped in their closed system of mutual back-scratching. We need to replace the system. If tomorrow we all listed our homes on craigslist or an equivalent, we would pull the rug out from under the MLS. Some real-estate agents—the smart ones—list homes in these alternative databases today. Shoppers may also list their desires to buy or rent homes or find roommates (as happens on craiglist), and someone—say, Google—could write an algorithm to link seekers and sellers directly, making the internet itself the marketplace. Other services feed the market with information it needs to be efficient. Zillow.com, for example, collects recent home sales so both buyer and seller can judge for themselves what a fair price would be.

As soon as the first real-estate agent (or agent's husband, as often happens) reads this chapter, I know I'll get an angry email or blog comment telling me I just don't understand the value they bring. But if you must explain your value, it's not as great as you think. With all due respect, that reaction betrays the same defensive, protective thinking torpedoing other industries covered in this book. The wiser reaction to such a challenge would be to see the opportunity in it. I'm not necessarily out to destroy agents. I want to wake them up. If you're the smartest, most competitive agent around, you should want to leapfrog your cozy competitors, disrupt their businesses, and exploit the new opportunities online brings. Or a newcomer will.

Sellers and buyers still need services. Perhaps the next-generation agent should offer them à la carte. First, sellers want buyers to find their homes. That's marketing. Agents say that's what they offer now, but they don't much. As I said earlier, when agents put an ad in the paper it's to market themselves as much as the home. I'd start a company that does nothing but help market homes in the open internet, creating listings on craigslist, taking pictures and making videos, making web pages for the homes, making sure those pages show up in searches, even buying ads on Google. Thanks to Google, you can do this on your own with links to as many photos as you want (free on Google Picasa); video tours (free on YouTube and easily shot with a $100 Flip Video camera); maps to area attractions (free with Google Maps); an aerial view (thanks to Google Earth); and lists and reviews of local restaurants (thanks to Yelp, also on Google Maps). Home sellers can add links to their own favorite hangouts and best grocery stores and add tips about where the kids can play. You can sell not just the property but the experience, the lifestyle, the community. It won't be long before you can introduce buyers to our neighbors, linking to their blogs or Facebook pages. Many homeowners wouldn't want to do this themselves, so there's a business opportunity to help. I'd sell these services and options for flat fees, not a percentage of the sale price.

The other problem with selling a house is hassling with tours. I'd start a company that offers concierge services to schedule and escort would-be buyers. The concierge doesn't have to sell the house (as a buyer, I don't need anyone to open closets and point out how allegedly large they are, thank you very much). Buyers could pay the concierge to chauffeur them from home to home. Sellers could pay the concierge to hold open-houses (and make coffee and cookies)—and I think that if buyers knew they wouldn't be trailed by sellers' agents, they might be more likely to visit a home. I would not be surprised to see local home-tour bloggers emerge, taking tours, taking pictures, and treating new homes on the market as news. I'd read it and buy ads there.

Closing is the final hassle. We need to change laws to simplify the process and shift control and advantage from lawyers to home buyers and sellers.

There are also technology opportunities. I'd like a mobile-phone application tied to Google Maps and global positioning so, as a shopper, I could enter my requirements—houses this large in this price range in this area—and the phone could map out a day's house hunting, scheduling the day and giving me directions so I get to open houses at the right hour. The application could show me photos and videos. It could contact concierges, agents, or sellers to make appointments. Who wants to drive around with an agent in a high-mileage Mercedes when you can go it alone? Or maybe I'm just being antisocial.

Buyers can use the tools of the web and mobile technology to research a prospective neighborhood. New services such as EveryBlock.com list all kinds of data around addresses—crimes, building permits, even graffiti cleanings. Outside.in organizes local blog posts around locations so you can read what your neighbors are talking about. With smart searches, home buyers can get school data and local news archives. They can look up and contact Facebook users who live in the area. A neat new service called CleverCommute provides a real picture of traffic headaches. All this open data beats the agent telling you that every neighborhood is wonderful and every house has potential.

Agent 2.0 will have her own rich web site showing the towns she covers and the homes she has helped to sell, with links to lots of information about the area. She'll want Googlejuice. When I come looking for a home, I may search for someone to help me. That could be a remade agent, it could be a disruptive newcomer, or I could do it on my own. I'll be looking for the best service and the best deal in an open and competitive market—without anyone paying 6 percent.

Google Capital: Money makes networks

I can't think of a Googlier industry than venture capital, and that stands to reason: Venture capitalists traffic in innovation, change, and risk. They watch what Google does, covet its success, and follow its investors. When I told venture capitalist Fred Wilson, a partner at Union Square Ventures in New York, that I was writing a book called *What Would Google Do?* he smiled and said, "We ask that all day long. That's our investment strategy." He and his partners also ask, "What would Sequoia do?" Sequoia Capital backed Google. That's Google envy.

Wilson is the Googliest guy I know in this, the Googliest industry. He was one of the first VCs to blog. When he started, his competitors thought he was nuts. Venture capital was a secretive business. You didn't want adversaries to know what you were thinking or the trends, companies, and people you were tracking. The goal was to get there first; it's a race. But Wilson benefited from revealing his thinking publicly. It helped him hone his ideas and attracted deals—one third of his investments come from the blog and online conversation, he says. Because of his publicness, Wilson developed a reputation online and a wider network of acquaintances who could help him do his work. When I saw him, he was about to head off for a month in Europe, where he wanted to find businesses at the far reaches—Slovenia, for example. He mentioned his trip at the end of a blog post and instantly had 100 people all over Europe wanting to meet him. As he traveled, I followed his meetings via Twitter updates.

"Being public and searchable and findable," he said, "is an important piece of it—owning the first page of your Google search, getting the brand of Fred Wilson out there." When you search Google for Fred Wilson, the first result is his blog (at avc.com); the second and fourth are pages with his bio; the fifth is his Wikipedia entry; the sixth is his Tumblr blog (on a platform created by a company he invested in); the ninth is his Twitter feed (another of his investments). There are other Fred Wilsons on the page: an artist who has had PBS documentaries made about him and a band by the same name. But according to Google, I was talking with *the* Fred Wilson.

On his blog, Wilson gets to try out ideas and products using these new platforms and tools. He has driven readers crazy cluttering his blog page with too many cool new widgets. But then he invested in many of those tools. Wilson's attitude about his blog and investing resembles mine about my blog and media: We learn, experiment, extend our reputations, and meet people. He uses his blog to help run his business; he found his latest associate through a blog post. He advises other companies to hire "net natives" who understand the new world because they live in it—and there's no better place to find them than on the net. Wilson inspired many of his competitors—mostly those who invest in the web, not in other big-iron arenas such as biotech and technology infrastructure—to start blogging. Now a score of prominent VC bloggers write posts explaining to entrepreneurs how to pitch VCs and how to run companies.

This ethic of sharing carries over to the companies in Union Square's portfolio. Wilson told me that one of his investments, Clickable, a search-engine marketing company, joins discussions on other sites just to answer questions that often have nothing to do with the company. They don't necessarily promote Clickable. They share knowledge like good citizens of the gift economy. "Their trail is their brand," he said. He told me about the head of another start-up who relishes getting into conversations—even with users who are angry when his service gets overloaded—because he learns so much about what users want.

Web 2.0 platforms—open and inexpensive software and services that make it easy and cheap to start new sites, services, products, and companies—present both opportunities and challenges for investors. The law that says small is the new big can make life hard when you are accustomed to making big bets, as VCs do—because they also want big returns. Today, a lot of new companies simply don't need VCs' money and when they do, they need less. If VCs have to invest in smaller increments in more companies, it is harder for them to manage their portfolios, which increases the cost and risk of investing. Never thought you'd feel sorry for a VC, did you?

Consider Outside.in, a company started by author, journalist, and entrepreneur Steven Johnson. Outside.in organizes local blog posts and their conversations around places and topics. It makes ingenious use of Google Maps, free databases, and other open-source software. Johnson launched the service using only $60,000 from an angel investor. If he had tried to build the business even five years before, he said, it would have cost $50 million. He could not have afforded, for example, to create the mapping technology Google gave him for free. To expand and hire staff, Outside.in took investment from venture capitalists, including Wilson's fund, but that amounted to nothing near $50 million.

As investments get smaller, entrepreneurs are also getting younger. Many of web 2.0's explosive new companies—Facebook and Digg, to name two—were started by people in their 20s. "The most interesting things I've seen this month and this year are the creations of kids who barely shave," Wilson blogged. This, he argued, is no accident. "It is incredibly hard to think of new paradigms when you've grown up reading the newspaper every morning. But we have a generation coming of age right now that has never relied on newspapers, TV, and magazines for their information and entertainment. They are the net natives. They grew up in AOL chatrooms, IMing with their friends for hours after dinner, and went to school with a Facebook login. The internet is their medium and they are showing us how it needs to be used." They are helping to build what the internet is becoming, which is what Wilson wants to invest in.

That blog post irked a bunch of entrepreneurs my age (hint: my beard is gray). But Clay Shirky defended Wilson's thesis on youth, arguing, "The principal asset a young tech entrepreneur has is that they don't know a lot of things. In almost every other circumstance, this would be a disadvantage, but not here, and not now. . . . When the world really has changed overnight, when wild new things are possible if you don't have any sense of how things used to be, then it is the people who got here five minutes ago who understand that new possibility, and they understand it precisely because, to them, it isn't new."

Shirky speaks for my generation when he says he knows from experience that you find music in stores, try on pants before you buy them, and get news and jobs reading newspapers. "I've had to unlearn every one of those things and a million others. This makes me a not-bad analyst, because I have to explain new technology to myself first—I'm too old to understand it natively. But it makes me a lousy entrepreneur."

Wilson responded to the fuss saying that he was not an ageist, only that he and his partners were seeing more and more young people with new ideas. "This is 15- to 20-year-old kids building and launching authentic web services that fill a real need in the market." As he blogged that, he linked to a web site run by my son, Jake, who was 15 at the time and had just written and sold a few Facebook applications, one of them to another venture firm. On a trip to Union Square's offices, after Wilson and his partners quizzed Jake about his net-native worldview, he advised Jake to find a technology mentor and suggested David Karp, who created the tool Tumblr (a Union Square investment). Wilson warned me that Karp had left high school to start his company. *High school.* (My wife to our son: "Don't you even *think* of it!")

How do investors meet entrepreneurs from a different generation? I think they need to operate in more open networks with more stakeholders at the edges. VCs are still a chokepoint of control: They raise money from investors; they pick and manage relationships with start-ups; they pay investors and keep their share. They are middlemen and Google makes detours around middlemen. As VCs are stretched thin—making more and smaller investments—it's harder for them to stay in the middle. It's also harder to find and evaluate new companies. I get a headache reading the popular blog TechCrunch, which covers new web 2.0 companies, because I can't hope to keep track of them all: mobile companies, social companies, companies dedicated just to managing blog comments. The low cost of launching and running new enterprises means they can serve niche needs in a small-is-the-new-big age. But the barrier to entry to competitors is also about a millimeter off the ground. It's harder than ever to figure out which of many competitors in a space will win.

So investors need to use a wider network of trusted people to help find and then manage new companies. Taking investment capital from these trusted agents and giving them a share of the profits if their finds pay off could form a network of miniVCs backed by the bigger VC. A variant of this model is New York Angels, a group of 65 successful investors who judge early-stage companies together. Incubators take a more active role in getting companies off the ground. Holtzbrinck, a publishing conglomerate based in Germany, runs a lab that starts some companies and invests in others, then decides whether to buy them. Idealab, founded by nonstop entrepreneur Bill Gross, has launched a large number of companies as an incubator, including Overture (which became the basis for Yahoo's—and, indirectly, Google's—search-ad industry), PetSmart, Picasa (now Google's photo software), Citysearch, and the electric-car company Aptera Motors. Both incubators provide space, office services, advice, and money. Then there is a series of next-generation incubators built to advise and invest in new web 2.0 enterprises. These include Y Combinator, which funds small entrepreneurs and helps them get from idea to company; Seed Camp, which runs regular competitions for start-up help in Europe; and Betaworks, which funds and advises early start-ups.

Investors still need to reach into the dorms at MIT and Stanford—or farther back into my son's high school—where ideas are hatching. I decided to teach because I was no longer able to effect enough change in a media company and figured I could do more in the cause of innovation helping students as inventors. At the City University of New York, I started a class in entrepreneurial journalism to prove that's not an oxymoron and to teach journalists business. My students create business plans for sustainable journalistic enterprises and, thanks to a grant from the McCormick Foundation, the class funds the best of them with seed money. Underlining Wilson's observation about age, my students do best when they think like young people. They fail when they try to think like graybeards. It is sometimes the graybeards who point this out to them. Jim Kennedy, head of strategy for the Associated Press, heard all my students' presentations and then told them he was disappointed that they had all proposed web sites. He said "web site" practically with disdain, as one would say "disco." He inspired one student, who wanted to start an online magazine for teen girls, to shift from the web to Facebook. She had to think differently.

Entrepreneurship is spreading among youth. There's a blog for young capitalists called College-Startup.com (tagline: "Get rich from your dorm room"). A 2007 Harris Interactive survey on entrepreneurship commissioned by the Ewing Marion Kauffman Foundation found that 63 percent of youths between 8 and 21 years old said they had the ability and desire to launch businesses, and 40 percent planned to do it. Thank goodness for the arrogance of youth.

Perhaps venture capitalism should start to look like a classroom: VCs could provide not just funding but also education (which some do in their blogs). If I were a VC, I'd reach out to colleges and offer to help talented entrepreneurs, dangling seed money for those with great ideas. I might open my doors as an incubator and offer free help to great business ideas so I could invest in some of them. (We will discuss other ways to nurture innovation in colleges in the chapter, "Google U.")

Or perhaps venture capital could look more like an open marketplace. When I asked mega-entrepreneur Gross how he'd make his field Googlier, he said: "I've always thought there should be a better start-up marketplace, almost like a mini stock exchange for start-ups, open only to qualified investors. But open up all the information and make it more even and transparent." The problem for founders and employees is that they can't take money off the table, as the saying goes, until a company is sold or goes public. A high-end marketplace of private start-up equity would let them sell a little stock to buy their Beemers but still keep working. Facebook did that in 2008 by enabling employees to sell stock to each other. Gross, the entrepreneur's entrepreneur, would like to start a company to build such a marketplace.

Can large companies spark entrepreneurism in their ranks? Google, of course, invests in internal innovation through its 20 percent rule. Google also buys innovation when it acquires companies. Apparently that hasn't been sufficient, for Google surprised the investment community when it started its own venture fund in 2008. When I worked in large companies, I saw how hard it was for them to invest in start-ups. Investing requires different skills. Finding start-ups comes from networking. Managing the relationship is more like teaching. And big companies need the patience to let investments grow on their own paths and timetables. Still, supporting innovation is vital in any industry and any company you can name. Rather than implementing 20 percent rules, perhaps companies can find innovators within their ranks by offering grants to entrepreneurial employees in return for equity. Perhaps universities can help. I am working to start an incubator for the news industry inside my university.

Venture capital's goal is to find talented people with good ideas and to give them the resources they need to execute those ideas. If I were a venture capitalist trying to think like Google, I'd figure out how to build a platform for entrepreneurship. I'd be as open as Fred Wilson with my ideas and hope to attract many more. I'd consider being a matchmaker more than a middleman, sometimes connecting investors and start-ups directly and trying not to get too much in the way. I'd rely on a large, distributed network of trust to help me find and manage investments, rewarding people in that network. I'd put together networks of start-ups that could help each other, whether I invested in them or not. I'd assume that just as it has gotten easier and cheaper to create content and media, it will get easier to create other kinds of companies. I'd manage abundance. All that, of course, assumes that I have an abundance of money. I don't. Oh, well.

The First Bank of Google: Markets minus middlemen

Banking is the ultimate middleman business, pooling money and need and profiting on the connections. In small ways—as in small is the new big—the internet is already disintermediating the industry by making direct connections.

Take peer-to-peer lending. At Prosper.com, as of 2008, 750,000 members had borrowed and lent more than $150 million in amounts as small as $50, supporting anything from launching new businesses to paying off college loans to getting out from under credit card debt. It's wonderfully simple and magnificently human. You see the person and the story: "It has been my dream to open a Neapolitan Pizzeria ever since I moved to the United States 9 years ago. I decided to start small at first and open a small food cart, and expand from there. . . . It is time we expand our little food cart to a full-size pizzeria." I wanted to invest in that one. Another: "This loan will be used to start a part-time business doing cooking classes for Raw Foods." What? Cooking raw food? I thought I'd pass on that. Then there was a student who wanted help to pay for her last year in college. "I work a full time job as well as go to school. I currently have a GPA of 3.9 overall. I am obtaining my degree in Accounting and Financial Management and understand the importance of paying your bills on time and maintaining a good credit score." OK, she sold me. Prosper advises users to diversify loans in case some default (it sends out the collection agency). Though the interest rates run high, this is no way to get rich or to build the new Bank of America. But it is compelling and entertaining. Prosper turns the most impersonal industry there is into a real-life reality show filled with dreams and winners and losers.

Other variations on the theme: Zopa sells certificates of deposit and gives investors a voice in lending the money. Loanio is supposed to make loans safer by involving cosigners and getting more documentation. Virgin Money handles the details in loans among friends and family. Lending Club makes loans social via Facebook.

Microloans are better-known—and tend to go to better causes—in developing economies, where they are used to buy cattle to start a business or to send a child to school. Go to dhanaX.com to see stories in India—where only Indians may invest—or to Kiva.org, where you can grant loans as small as $25 to businesses all over the world. Kiva has made $35 million in loans—averaging $485 each—in 43 countries, and 98.1 percent have been paid back. (For comparison, 2.7 percent of prime loans in the United States were in default in the spring of 2008; subprime loans saw 16.6 percent in default.) One Kiva request: Mrs. Phally of Cambodia raises pigs at her home, making $7 a day, while her husband farms, earning $5 per day. The family is also supported by two of their children who work in the local garment factory. Mrs. Phally asked for a $1,000 loan to buy a small tractor for her husband to plow his land. Adding context, Kiva tells us it is common for Cambodians to rent out tractors to make extra income. Kiva's loans earn no interest for the lender, but local administrators charge interest. These loans change the world one entrepreneur at a time. That is an internet dividend.

To make a similar impact in the United States, a bit at a time, see DonorsChoose.org, where you can contribute to teachers' needs. See also Facebook's Causes application, where members start, join, support, and donate to causes. All these new entities rely on small bits adding up to big impact, on direct and personal connections, on giving control of the use of resources to those who have them, and on open information.

The root of the credit crisis that spread from America around the globe in 2008 was that bad loans were hidden in packages with good loans and sold to financial markets, with no accountability down to the level of each loan and no transparency. That's not the case in these peer-to-peer loan operations. I don't mean to pretend that the social banking system could replace banks, but banks could learn a lot from it. Why not set up direct marketplaces that let me establish my own diversified portfolio in small-business loans, home mortgages, and student loans? Why not use the infrastructure the bank has, as Virgin Money and PayPal do, to facilitate our own financial transactions? Why not make banks human again? We may not see such an evolution in big, old banks—they're just too big and old. That is why we are seeing new and innovative, peer-to-peer banks and financial institutions emerge. But there can be no question that the industry needs both more transparency and more accountability.

The internet also presents new opportunities for financial markets. Online we have new sources of information and analysis about companies other than the conflicted analysts who work for financial institutions. Investors themselves can share knowledge, data, strategies, successes, and failures. The Motley Fool's CAPS service pools investors' knowledge to help each member of the community. I invested in Covestor, where stock investors share their verified trading history and others will be able to invest alongside them. Any investor can become his own mutual fund and a winning investor has another way to benefit from betting well.

In my entrepreneurial journalism class, Shirky advised students working on a personal finance site to offer a branded credit card, enabling them to aggregate data from the community to let people know where they stood against peers: "Warning—you are spending 15 percent more on restaurants than people your age with your income." Learning from lots of data is a pillar of Googlethink. Banks and credit cards know more about our spending than anyone and almost as much about our buying as Amazon. That's our data as individuals and our wisdom as crowds. I wish they'd turn it over to us so we could use what we learn from it to manage our finances.

Of course, banking and financial markets are regulated for good reason—not closely enough, judging by the results of the credit crash. We need to tread with caution in these areas. But the web presents new ways to think and do business, even in the stodgy old business of handling money.

I'm surprised the web hasn't had a greater impact on the industry already. Every time I see a new retail space being built in my area, I get depressed when I see a bank move in. How useless. How unfun. I'd prefer another Starbucks or a Taco Bell or maybe an old-fashioned bakery. Why are banks still in the business of building so many retail outlets out of bricks and filling them with staff? When the internet arrived, so did some online-only banks, but they never flourished and many were bought up: in the U.K., Egg was acquired by Citi, First Direct by HSBC. They didn't offer us enough incentive to change our habits. If online banks had passed savings onto us—the internet dividend in cash—maybe we'd have been motivated to go virtual.

The cashless society will probably come to the U.S. a day after the paperless office does—that is, never. We keep hearing about people in Finland and Japan buying Cokes and paying for parking with their mobile phones, but we haven't seen that happen in the States. Microsoft wanted to become the cash register of the web with its Passport service, but I think no one trusted Microsoft to handle our money. Google's Checkout service has not caught on. PayPal, now owned by eBay, has become an easy way for people to exchange cash, but too few merchants use it. Maybe we need a new virtual currency all the world could share that could become the basis of new financial systems. How does Googlebucks sound? In Google we trust.

Public Welfare

St. Google's Hospital
Google Mutual Insurance

St. Google's Hospital:
The benefits of publicness

Too often when I find myself in a discussion about citizen journalists, some member of the press' curmudgeonly class—thinking himself quite clever and apparently believing he just thought of this himself—will growl at me: "Why should I trust a citizen journalist? You wouldn't want a citizen surgeon, would you?" No, I wouldn't.

But I do want health care to open up to the Google age and take full advantage of the opportunities it presents to gather and share more data; to link patients with better treatment and information; to connect them with fellow patients in a community of shared experience and need; and to use the potential of collaborative tools and the open-source movement to advance medical science.

On my blog, I have violated the most sacred tenant of privacy advocates: I revealed and discussed my personal health information, writing about my bouts of atrial fibrillation (a sometimes irregular heart beat—I'm fine, thanks). I have received great benefit from opening my medical history to my readers. Fellow patients have given me support, sent me links to resources, shared their experiences about treatments I've considered, and sent me updates on companies working on new treatments. Even Google's Sergey Brin blogged that he had learned he carried a gene mutation that may indicate a propensity for Parkinson's disease.

Imagine how valuable it could be for us patients to go to a site to record our conditions and activities right before the onset of afib (the familiar name for the condition). In some people, too much food, wine, stress, or activity can trigger an attack; in others, these have no effect. Doctors have some of this data already, but only from limited samples. If millions of patients around the world shared their experiences, would we discover new triggers, new correlations, new causes, even new treatments? Don't know. But we can't know until we try, until we open up and provide the means to gather the information and analyze it.

PatientsLikeMe has created a platform for communities around a still-limited set of conditions, including multiple sclerosis, Parkinson's disease, depression, and post-traumatic stress disorder. I spoke with the husband of a woman who, months before, had been diagnosed with MS. He said the site has been invaluable, providing information, experience, and support. The 7,000 MS patients in the group—growing by more than 700 a month—categorize themselves by symptom and treatment and submit narratives and quantitative data: We can see that 395 patients took a particular drug for fatigue; 23 stopped taking it because the side effects were too severe, 21 because it didn't seem to work, and 14 because it was too expensive. This experiential data is a goldmine to a patient trying to learn about and take greater control of her treatment. It is also valuable to the medical industry. The company explains that its operating costs are covered by "partnerships with health-care providers that use anonymized data from and permission-based access to the PatientsLikeMe community to drive treatment research and improve medical care." When we share information in a network, all its members may benefit.

To build these networks, we need to think of health as a public story and rethink certain inhibitions to publicness. We are not motivated to be open when insurers or employers can reject us due to preexisting conditions. Not that I want to push a political agenda, but universal health care would solve much of that problem. Even then, I'm not suggesting that everyone reveal all their ailments. I understand if you don't want to talk about yours. But there could be benefits if you do. Health is just one illustration of how the internet's ethic of publicness could have a subtle but profound impact on how we live.

In 2008, Google started a health service online (at google.com/health) where users can enter their conditions and the drugs they take as well as results of tests, such as cholesterol screening, which they may download from a limited number of health companies that have signed up so far. Patients' information is not meant to be public, though a few of us online folks have wished that we could publish our own pages openly so we could reap the benefits of medical networks. Google's purpose is to give users more information (it feeds me news stories about afib) and to put users in control of their own information, because they have too little control now.

There is a movement afoot to standardize personal health records. It is related to another movement to create systems where customers control relationships with vendors—called vendor relationship management (VRM), the mirror image of customer relationship management (CRM). VRM is being spearheaded by pioneering blogger Doc Searls, a fellow at Harvard's Berkman Center for Internet and Society. I view what he's doing, shifting control to customers, as Jarvis' First Law brought to life. Searls, who's not an M.D., turned his VRM attention to health after spending a tortuous week in the hospital with pancreatitis, which he chronicled from his bed on Twitter and then on a blog. He complained about the lack of information he had, which led him and his doctors to ill-informed decisions that exacerbated his condition. "I believe the closed and proprietary nature of health care is itself a disease that needs to be cured," Searls said as he linked to another blogger, Fred Trotter, who illustrated the problem of getting control of our own health information. "Let's imagine that I had some kind of life event that would require me to gather those records together," Trotter blogged. "To do that, I would need to call every doctor I have ever visited, and request a copy of my records." Those doctors would all want to fax him records. "Faxing over phone lines is the 'health exchange network' that we have in the United States," Trotter said. He would end up with a giant pile of documents that is not searchable, is hugely redundant, and is not easily read. His doctor is unlikely to spend the time needed to sift through it all looking for that nugget of a clue.

Searls argues for open standards in medical information to organize data and put it under patients' control. He compares the task to the creation of the internet itself (or as I would put it, bringing Googlethink to medicine). "We cannot fix health care only at the institutional level," he blogged. "No company and no government agency can fix health care, any more than any company or government could fix networking or computing. Those had to be fixed by hackers building solutions for everybody and not just themselves." Searls hopes to look back in his lifetime and see that health care was reformed from the bottom up thanks to the open-source infrastructure of the internet. He also hopes to see new businesses created "atop patients as platforms."

Now apply this attitude—this ethic of openness, standards, and hacking—not just to medical care but also to medical research. How much of pharmaceutical work would benefit if more data were open and more of the work were open-source? We've heard the arguments: The cost of developing drugs is astounding and unless companies that create them can fully own the information and the results and recoup the expense, they won't discover the next pill that could save your life. I don't disagree; I respect their work, their business needs, and their intellectual property. Still, we need more discussion on the impact openness could have on medical research. Would the government need to sponsor more research so the results would be open? If universities, governments, and doctors shared their data in standard, open, and free databases—with patients encouraged to add their knowledge and experience—would that have a greater benefit than the current, less-transparent structure? If more research were made open, what drugs and businesses could result? Who could organize that knowledge for us? Google has opened up most human knowledge today—any that is digital and searchable—so I'm confident it could do the same with medical knowledge. Like Searls, I hope to live to see that day.

Medicine is still too much a priesthood of closed knowledge, at least as it relates to patients. In 2008, I sat with doctors from around the world at a conference lunch as they clucked, scowled, and shook their heads and shared stories of their patients going to the internet and coming back with incomplete or mistaken information. These doctors wished that their patients hadn't done their own research and that the doctors, as experts, could have kept control of access to information. Well, too late. I advised them to curate good information for their patients. What if they created resource sites? What if they blogged to keep patients informed and up-to-date—and also linked themselves with a larger community of doctors working on the same conditions? If their patients got more of the right information, would that make them better patients? A bit grudgingly, the doctors accepted the notion. I've debated my prescriptions and treatments for afib with my doctor and what I really want from him is data and information about my choices to make better decisions together. I'm no citizen cardiologist, but it is my heart.

I want more information to be made public about doctors as well. It is possible to get survival rates for hospitals performing certain procedures (though sadly, keeping these scores sometimes disincentivizes institutions from taking hard cases). Patients rate doctors—like teachers and plumbers—at various online services, but they're not terribly helpful because I don't know anything about the people leaving comments. I'd at least like to get a list of all the conditions a doctor treats and how often so I can pick the most experienced specialist. If a Googley restaurant would tell me how many diners ordered the crab cakes, a Googley doctor should tell me how often she has treated afib. I would also be impressed if the doctor treating me had written about the condition online. I'd be doubly impressed if I saw other doctors linking to her.

The changes in medicine we've touched on all relate to information: opening it up, sharing it, organizing it, analyzing it, bringing the network effect to the industry and our health. That is Google's specialty.

Google Mutual Insurance: The business of cooperation

As I was researching this book, I wrote on my blog that I had come up against a few industries I thought were immune from reform through Googlification. Insurance topped my list (we'll get to the others shortly).

Insurance is built on getting us to take a sucker bet—a bet even we want to lose. Nobody wants a reason to collect collision, fire, flood, health, or certainly life insurance. Worse than Vegas, we know that insurance companies stack the deck against us; that is the foundation of their business. If we don't collect, we are losers (we've lost our money). If we do collect, we're still losers (something bad happened). If the insurance company pays out too much and goes out of business, then those of us who paid in still lose. We can't win. The industry has to suspect that we are liars, making us prove our misfortune and reluctantly giving us back the money we put in the pool. They make the economics overcomplicated so we don't know just what suckers we are and so we keep making safe bets—safe for the insurance company. Our relationship with insurance is, therefore, necessarily adversarial and built on mutual mistrust. How incurably unGoogley.

My readers disagreed. A few dozen of them left comments on my blog arguing that insurance can reform, and they showed me how. Here are excerpts from the conversation and my education. (Let this chapter be an object lesson in the power of open, collaborative thinking.)

The first comment came from Seth Godin, author of *Purple Cow, Small Is the New Big, Tribes,* and other business best sellers, who scolded me: "Think bigger, Jeff!" He provided a few examples of social insurance. First:

> 20 Korean families pool finances and open businesses one at a time . . . each member of the group has a huge incentive to help each business succeed, so they can get the money when it's their turn. Imagine insurance being created in a coordinated fashion, with each member of the coop working to decrease the risk of everyone in the pool.

A commenter from France, Bertil Hatt, said the Mutuelle Assurance Instituteur France (MAIF) lives by some of these principles of mutual benefit, providing insurance as well as services, such as home and child care. Premiums are higher than average, she said, but lower for the young, the poor, and students. "How can they make it?" she asked. "Thanks to an implicit contract: When you get richer, you stay with them not only for the service, but because you believe in their way." Insurance becomes a collective, though private, good.

Godin next talked about smart devices that might need less insurance. Cars with better brakes can cost less to insure if they keep us safer and also cost less to repair and to warranty—which, again, is a form of insurance. Godin took the idea a step farther and suggested that "smart products come with their own insurance because they're so much better and talk to each other."

When cars know where they are and where trouble might be, or
when they integrate with each other and their drivers and the roads
and the police, shouldn't insurance get better?

Right. The network becomes a form of insurance as connected devices
can be monitored, repaired, and improved and can learn to do their jobs
better and more safely. In the comments, Chris Cranley took off on Godin's idea and suggested that just as smarter products may need less insurance, the same may be said of smarter people: "If I knew how to avoid
problem X, I would not insure against it." Education and information
become insurance against insurance. Godin took this line of thinking to
its extreme when he speculated about opportunities not just for smarter
people but—genetically speaking—healthier people as determined by
23andMe, a service that analyzes users' DNA. (Founded by Brin's wife,
Anne Wojcicki, 23andMe discovered his Parkinson's gene. Google invested
in the company.) Godin said:

And while some may not like it, what happens when 23andMe gets
a lot smarter and the healthiest gene pool starts their own life insurance coop?

U.K. business journalist James Ball agreed with me that insurance is "a
glorified betting market" where insurance providers "offer odds against
certain outcomes—adverse outcomes—and we pay up the stake. The
similarity between insurance providers and bookmakers stack up easily."
His comment added that open betting exchanges had shattered bookmakers' control over odds and premiums and could do the same in insurance. "There's no built-in reason for 'social' insurance to fail," he wrote.
"In fact, it could work quite well." Ball was arguing, as I have many times
in this book, that the power of open information will make markets more
efficient. He gave me a dose of my own medicine and I had to agree.

But still, I argued in response, there is the issue of fraud: People try to
rip off insurance companies and that can undercut communities and markets built on trust. Ball replied that fraud is less of a problem in some
cases. "Let's suppose we have insurance against burglary by requiring the
crime to be reported before paying out," he wrote. That requirement gives
insurers a measure of security. He continued:

Risky, or trusting, insurers could offer worse "odds" with less requirement for proof in the event of a claim. By treating insurance
as any other betting market, we'd effectively be insured by many
small stakeholders.

Cleverer yet, the marketplace could take a cut of premiums in
some markets (say 5 per cent?) and use this to audit a random percentage of claims, for particularly risk-averse insurers, or for markets particularly sensitive to moral hazard.

Ball said his insurance marketplace would use technology and the
theory of social networks to rely on transparency more than trust. He
concluded: "Health insurance would certainly take some thought. But
then again, I'm in the U.K., so not a problem for me." Rub it in, why don't
you, James?

Shaun Abrahamson, a friend and former colleague, piped in to the
comments, pointing out that the original insurance companies, like credit
unions, "would have been recognizable as precursors to social networks."
Then he pushed the social envelope in the discussion: "To James' point
on betting and odds-making, do you think groups of people who know
one another might outperform actuaries in assessing risk? Do you think it
would be easy to defraud a network of people who know each other via
friend-of-a-friend type connections?" In other words, if a community in-
sures itself, are there social disincentives against screwing friends and
neighbors?

Ivan Pope, a U.K. web entrepreneur, echoed Abrahamson and told me
my premise was wrong. Insurance, he said, is inherently social.

In the same way that mutual societies and co-operative societies
are all social, so insurance is a social contract. We all put in a bit
and the ones who need it draw down from the pool. Sure, we priva-
tised the management of it, gave away the profit, turned it into a
huge scale business. . . . So we need some imagination, some
ambition and some skill to build these back again as social communities.

Scott Heiferman, founder of Meetup, also brought historical perspective to the discussion, writing a brief manifesto for change in the coming

decade, chock full of hip blog references (the "social graph" to which he refers is what Mark Zuckerberg calls the architecture of personal connections on Facebook):

> Historically, when people are free to assemble & associate, they self-organize insurance, cooperatively. Later it became the centralized, professionalized industry we know today. I predict there'll be some kind of massive craigslistification of insurance by April 27, 2018. It's about de-institutionalization—not from the government borg (social security), not from the corporate borg (AIG). The New Social [graph] Security. Decentralized, self-organized. Not just DIY, but DIO (Do It Ourselves). That is the big theme for everything now.

There is the great promise and power of the Google age: DIO.

In the end, commenter Gregory Lent summed up the ideal for the Google age, saying that the web 2.0 social network

> will blow up insurance, because it will transparently link the whole system, insured, insurers, providers of the service that insurance is paying for. no place to hide, accountability everywhere, prices will drop, profits/savings more evenly dispersed. best thing that can happen.

Tie all this together and we can begin to draw a picture of a disruptive insurance enterprise that empowers a community by handing over control of insurance to the members of that community. I played out this scenario for a couple of insurance executives who said I may be mad but the ideas are good.

Imagine a forward-looking company—Google, for example—creating a new insurance compact: If the community makes itself healthier and lowers the costs—and raises the effectiveness—of its own care, the cost of insurance will fall. The deal would motivate the community to pressure its members to become healthier and smarter. Insurance companies today try to get us to act healthier, pushing us to join health clubs or eat smarter. But—apart from our feeling better—the direct economic benefit on medical costs is almost entirely the insurance company's and we never see a transparent accounting of the impact. The insurance 2.0 compact puts the community in charge and gives it mutual benefit and responsibility. Giving the community control means giving it information. The insurance company would need to give members complete disclosure about actuarial data, costs, and profits. The insurance company would also need to pressure doctors to hand over data about their work so community members could make smarter decisions about treatment.

The community, in return, needs to manage its health care, including keeping an eye on health providers. For example, my medical group makes me come in every four weeks to get checked for the blood-thinning drug I take because of my afib. My results never vary. Every time I'm there, I'm amazed at the inefficiency I see: two nurses making a big show out of pricking a finger (which some diabetics do on their own a half-dozen times a day). The medical group profits from my copay and from my insurance company's fees. It's a waste. I'm not motivated to do much about it. I have no relationship with the insurance company except mutual distrust and inconvenience. I would get nothing out of protesting or whistleblowing. If my community and I were in charge of our health care and insurance, that would be different. I'd make it my business.

The community also might choose to sponsor races, diets, and classes and pay for that out of its pool of premiums if it believes the bet on health will pay off. It might offer services such as the French MAIF's home and child care if the group believes it is worth the cost. That becomes the community's decision. What emerges is a community whose members want to maintain better health at lower cost and risk through mutual benefit. They are able to do this because the new insurance company provides a platform with tools, information, and organization to help the community meet its goals. The insurance company's not in charge. The community is. It's a vision of insurance that follows many of Google's rules and starts with Jarvis' First Law.

This vision came from my readers. They applied the internet's new ways to old problems to see what could be improved. They believed that more transparency in marketplaces would yield greater value. They believed that adding social elements—the interests and pressures of a community—would increase value. They told me that handing control to the market would increase trust, and insurance is about trust. So they proposed networks of mutual need and service that diminish if not eliminate the middlemen.

I'm proud to say that I didn't come up with these ideas. My generous readers did. They were my insurance against an empty chapter.

Google U: Opening education

Who needs a university when we have Google? All the world's digital knowledge is available at a search. We can connect those who want to know with those who know. We can link students to the best teachers for them (who may be fellow students). We can find experts on any topic. Textbooks need no longer be petrified on pages but can link to information and discussion; they can be the products of collaboration, updated and corrected, answering questions and giving quizzes, even singing and dancing. There's no reason my children should be limited to the courses at one school; even now, they can get coursework online from no less than MIT and Stanford. And there's no reason that I, long out of college, shouldn't take those courses, too.

You may suspect that because I'm a professor, I'll now come out of this litany of opportunities with a rhetorical flip and demonstrate why we must preserve universities as they are. But I won't. Of course, I value the academy and its tradition and don't wish to destroy it. But just as every other institution examined in this book is facing fundamental challenges to its essence and existence in the Google age, so is education. Indeed, education is one of the institutions most deserving of disruption—and with the greatest opportunities to come of it.

Call me a utopian but I imagine a new educational ecology where students may take courses from anywhere and instructors may select any students, where courses are collaborative and public, where creativity is nurtured as Google nurtures it, where making mistakes well is valued over sameness and safety, where education continues long past age 21, where tests and degrees matter less than one's own portfolio of work, where the gift economy may turn anyone with knowledge into teachers, where the skills of research and reasoning and skepticism are valued over the skills of memorization and calculation, and where universities teach an abundance of knowledge to those who want it rather than manage a scarcity of seats in a class.

Who's to say that college is the only or even the best place to learn? Will Richardson, who teaches fellow educators how to use the internet in the classroom, wrote an open letter to his children, Tess and Tucker, on his blog, Webblog-ed.com: "I want you to know that you don't have to go to college if you don't want to, and that there are other avenues to achieving that future that may be more instructive, more meaningful, and more relevant than getting a degree." He said education may take them to classrooms and lead to certification but it also may involve learning through games, communities, and networks built around their interests. "Instead of the piece of paper on the wall that says you are an expert," he told his children, "you will have an array of products and experiences, reflections and conversations that show your expertise, show what you know, make it transparent. It will be comprised of a body of work and a network of learners that you will continually turn to over time, that will evolve as you evolve, and will capture your most important learning."

If that is what education looks like, what does a university look like? I asked that question on my blog and entrepreneur and technologist Bob Wyman (who works for Google) responded by abstracting the university and identifying its key roles: teaching, testing, and research. I'll add a fourth and unofficial role: socialization. Let's examine them in reverse order.

Socialization is, of course, a key reason we go to college and send our children there. Adults see college as a process of maturation and increased independence and responsibility. Students, on the other hand, may see it as a process of getting away from the parents. Whatever. Jeffrey Rayport, a consultant and Harvard Business School professor, sat with me in the Harvard Club in New York and told me it was designed by a graduate of the university who didn't much care for the school's harsh Cambridge atmosphere. In the club, he created what he wished Harvard had been:

warm wood and fires, Harry Potter without the pomp and kitsch, the experience—the Disney World—of education. I do think there is a time to have that experience and live with our peers. Old people do. My parents live in Sun City Center, Florida, a town where one legally may not reside if under the age of 55. Why not have youth towns where residents are evicted by age 30: Melrose Place University?

But seriously . . . if one has the luxury of time and resources to explore the world before buckling down to a job and a mortgage, great. That exploration can take the form of backpacking around Asia, hanging out in a dorm, or joining the Peace Corps. Or these days, it may mean starting a company. Our young years may be our most creative and productive. Bill Gates, Mark Zuckerberg, and the Google boys dropped out of school at various stages to start their corporate giants. Should we be forcing young people to go through 18, 16, or even 12 years of school—trying to get them all to think the same way—before they make things? Instead of the perennial call to subject our youth to mandatory national service—how's that for a way to squander a precious resource?—shouldn't we instead be helping them find and feed their muses?

Perhaps we need to separate youth from education. Education lasts forever. Youth is the time for exploration, maturation, socialization. We may want to create a preserve around youth—as Google does around its inventors—to nurture and challenge the young. What if we told students that, like Google engineers, they should take one day a week or one course a term or one year in college to create something: a company, a book, a song, a sculpture, an invention? School could act as an incubator, advising, pushing, and nurturing their ideas and effort. What would come of it? Great things and mediocre things. But it would force students to take greater responsibility for what they do and to break out of the straitjacket of uniformity. It would make them ask questions before they are told answers. It could reveal to them their own talents and needs. The skeptic will say that not every student is responsible enough or a self-starter. Perhaps. But how will we know students' capabilities unless we put them in the position to try? And why structure education for everyone around the lowest denominator of the few?

Another byproduct of a university's society is its network—its old-boy network, as we sexistly if accurately called it. That club has long held value for getting jobs, hiring, and making connections. But now that we have the greatest connection machine ever made—the internet—do we still need that old mechanism for connections? LinkedIn, Facebook, and other services enable us to create and organize extended networks (any friend of yours . . .) springing out of not just school but employment, conferences, introductions, even blogs. Members of Skull and Bones at Yale and graduates of Harvard Business School may object, but as an internet populist, I celebrate the idea that old networks could be eclipsed by new meritocracies. Facebook didn't just bring elegant organization to universities, it could supplant them as a creator of networks.

The next role of the university will be harder to nurture in a distributed architecture. Research, pure and directed, are values of the academe that the marketplace alone may not support. Unless it has a market value and is paid for by a company, research must be subsidized by foundations, endowments, donations, and tax dollars—and often by the generous passion of the researcher. That will still be the case. The question is whether research will be done in schools or in think tanks and whether it will be performed by professors or by paid thinkers. There's little reason that research must be performed on campuses by academics and little reason why those academics cannot work in wider networks. Research has long been a process more than a product as papers are peer reviewed and research results are replicated. That is even more the case now as research is opened up online in web sites, blogs, and wikis and as their contents are linkable and searchable via Google (which provides a search service for academic works at scholar.Google.com). This openness invites contributions, collaboration, and checks.

The next role of the university is testing and certification: the granting of degrees and anointing of experts. The idea of a once-in-a-lifetime, one-size-fits-many certification of education—the diploma—looks more absurd as knowledge and needs change. Are there better measures of knowledge and thinking than a degree? Why should education stop at age 21? Diplomas become dated. Most of what I have done in my career has required me to learn new lessons—long past graduation—about technology, business, economics, sociology, science, education, law, and design. Lately I've learned many of these lessons in public, on my blog, with the help of my readers. That is why I urge other academics to blog and be challenged by their public. I believe that should count as publishing. Blog or perish, I say.

Our portfolios of work online, searchable by Google, become our new CVs. Neil McIntosh, an editor at the Guardian, blogged that when he interviews young candidates for online journalism jobs, he expects them to have a blog. "There's no excuse for a student journalist who wants to work online not to have one," he wrote. "Moreover, the quality of the blog really matters, because it lets me see how good someone is, unedited and entirely self-motivated." Our work—our collection of creations, opinions, curiosities, and company—says volumes about us. Before a job interview, what employer doesn't Google the candidate (a practice banned by law in Finland, by the way)? Our fear is that employers will find embarrassing, boozy pictures from spring break, but that's all the more reason to make sure they also find our blogs and collected works.

Sometimes employers will require certification. That, as Wyman says, is where testing comes in: exams to make sure our new doctors, lawyers, and PC support staffs know their stuff. But these exams are often given by professional organizations—medical boards and the bar—rather than schools. Preparation for those tests is undertaken by test-prep and commercial-education companies such as Kaplan. Universities ceded the market to them. Still, testing makes sense; it is our guarantee against the citizen surgeon (or that the citizen is qualified). It does make more sense to test students *after* they've learned a subject than before. Tests given *before* education commences—entrance exams—might better serve students if they discovered not what students know but rather what they need to know. Between SATs and exams mandated by No Child Left Behind laws in the U.S., we are succumbing to a tyranny of testing that commodifies learning. The system tries to turn out every student the same.

Finally we arrive at the core, the real value of a university: teaching. Here I violate my own first law when I say that complete control of one's education should not always belong to the student. For when we embark on learning, we often don't know what we don't know. Or in Google terms, we don't know what to search for. The teacher still has a role and value: If you want to learn how to fix a computer or operate on a knee or understand metaphysics, then you hand yourself over to a teacher who crafts a syllabus to guide your understanding. When it's clear what you want to learn—

how to edit a video with FinalCut, how to speak French—it's possible for a student to use books, videos, or experimentation to teach herself. The internet also makes it easy to connect teachers with students—see Teach-Street.com, which in only two cities has 55,000 teachers, trainers, tutors, coaches, and classes, according to Springwise. I wouldn't go there to learn surgery, but I might to get help with my stale German.

One benefit of the distributed, connected university is that students may select teachers. Instructors won't be able to rest on tenure (I speak as someone who has it) but must rise on merit. Today, instructors are graded on sites such as RateMyTeachers.com, but students are still prisoners to their school's faculty. If they could take courses from anywhere, a market-place of instruction would emerge that should lead the best to rise: the aggregated university. Instructors could also pick the best students. A class would become a handpicked team that might research a topic as a group, blog their collective process of discovery, or write a textbook and leave a trail of their frequently asked questions and answers for the next class or the public (what are courses but FAQs?). That product will be searchable and may provide a way for future students to find and judge courses and instructors. It's educational SEO, bringing the internet's ethic of transparency to the classroom.

There could be new models for education. One might be education by subscription: I subscribe to a teacher or institution and expect them to feed me new information, challenges, questions, and answers over years. Many schools give graduates refreshers and updates in skills; at the City University of New York Graduate School of Journalism, we call this offer our 100,000-mile guarantee. Education could be a club more than a class: We join to learn and teach together, sometimes handing the teaching duties to the best student on a given subject. Peer-to-peer education works well online as we can see in language-learning services such as Live-mocha, where teachers in one language become students in another and where anyone in its gift economy can critique and help any student. It is a learning network.

In the classroom, real or virtual, Google forces educators to teach differently. Why are we still teaching students to memorize facts when facts are available through search? Memorization is not as vital a discipline as fulfilling curiosity with research and reasoning when students recognize what they don't know, form questions, seek answers, and learn how to judge them and their sources. Internet and Google literacy should be taught to help students vet facts and judge reliability.

Is there a university, post-Google? Yes, these institutions are too big, rich, and valuable to fade away. But like every other institution in society, they should reshape themselves around new opportunities. Universities need to ask what value they add in educational transactions: qualifying teachers, helping students craft curricula, providing platforms for learning. We need to ask when and why it is necessary to be in the same room with fellow students and instructors. Classroom time is valuable but not always necessary. Many professional MBA programs have found ways to limit time together so that education need not interrupt life. The Berlin School of Creative Leadership (where I serve on the advisory board) has students meet in cities around the world so they can tap local expertise. Universities can become bigger than their campuses, and by bringing together special interests and needs from around the world, they can also become smaller, focusing on niches of knowledge while leaving other topics to other institutions. Schools, too, will do what they do best and link to the rest. That requires them to make their knowledge open and searchable; Google demands it.

How will universities work as a business? To quote former MIT professor and satirical songwriter Tom Lehrer about the famous German rocket engineer who came to NASA: "'Once the rockets are up, who cares where they come down / That's not my department,' says Wernher von Braun." If I taught three, three-credit courses a term for two terms to 20 students in each and they paid what they pay to my state-supported university—about $250 per credit—that would bring in $90,000, which is what I am paid (I don't do it for the money). In a competitive market, would students pay $750 for my class? That depends on the quality of my teaching, the reputation of the university, and the state of the competition. If they pay that amount, it still leaves no money for the university. Funds to support its structure would need to come, as they do now, from public or private subsidies. It doesn't look like a sustainable model.

Then again, look at University of Pheonix, Kaplan University, and other for-profit professional educational companies that have sprung up teaching students what they need to know for jobs. They're not academic like Oxford, but they fill a role and work as businesses. They charge more per credit-hour than my state institution but less than prestigious private universities. I think we'll see many entrepreneurial enterprises devoted to education emerge as the internet enables a new marketplace of learning. Perhaps different entities will maintain different roles. To learn database programming, you go to Kaplan; to learn the entrepreneurship needed to create a new Google, you go to Stanford.

On its official blog, Google gave advice to students, not about where they should learn but what they should learn. Jonathan Rosenberg, senior VP of product management, blogged that the company is looking for "non-routine problem-solving skills." His example: The routine way to solve the problem of checking spelling would be use a dictionary. The non-routine way is to watch all the corrections people make as they refine their queries and use that to suggest new spellings for words that aren't in any dictionary. Rosenberg said Google looks for people with five skills: analytical reasoning ("we start with data; that means we can talk about what we know, instead of what we think we know"); communication skills; willingness to experiment; playing in a team; passion and leadership. "In the real world," he said, "the tests are all open book, and your success is inexorably determined by the lessons you glean from the free market."

Rosenberg's best advice for students and universities: "It's easy to educate for the routine, and hard to educate for the novel." Google sprung from seeing the novel. Is our educational system preparing students to work for or create Googles? I wonder.

The United States of Google: Geeks rule

What if a Google guy were president? Earlier, I told of witnessing the competing worldviews of Larry Page and Sergey Brin versus that of Al Gore as they tackled environmental and energy crises. Google's founders saw the world and its problems through their engineers' eyes. Rather than seeking solutions through regulation and prohibition they relied on invention and investment: shouldn't do vs. can do. If the geeks take over—and they will—we could enter an era of scientific rationality in government. Other nonpoliticians have improved government. Michael Bloomberg ran New York City as a business. Arnold Schwarzenegger ruled California on the power of personality. A Google guy might just run government as a service to solve problems.

Whether or not they take charge, Google and the internet will have a profound impact on how government is run, on its relationship with us, and on our expectations of it. Now that we have the technological means to open up government and make every action transparent, we must insist on a new ethic of openness. So abolish the Freedom of Information Act and turn it inside out. Why should we have to ask for information from our government? The government should have to ask to keep it from us. Every action of government must be open, searchable, and linkable by default. The information government knows must be online with permanent addresses so we can link to it, discuss it, and download and analyze it. Government needs a new and transparent attitude: Officials and agencies should blog and engage in open conversations with constituents. They should webcast every meeting, since technology now makes that easy. Remember Weinberger's Corollary to Jarvis' First Law: There is an inverse relationship between control and trust. The more our leaders trust us with information, the more we will trust them with government. Right now, there's too little trust in both directions.

I want government to implement tools like MyStarbucksIdea and Dell IdeaStorm to enable citizens to make suggestions and share ideas, discussing them together as communities: GovernmentStorm. The United Kingdom has E-Petitions, a program launched by the prime minister's office in 2006 with help from citizen activists in mySociety, which creates tools for government openness. Among the petitions: "Scrap the planned vehicle tracking and road pricing policy" got 1.8 million signatures. "Cut value-added tax on 100% fruit juices and smoothies to the minimum 5% allowed by EU law to encourage shoppers to take the healthier option and achieve their 'five a day'" attracted 10,400. "Make breastfeeding in public legally acceptable for all babies and children" got almost 6,000. In its first year, 29,000 petitions were submitted (14,000 of them rejected because they were duplicates, jokes, or unlawful) drawing 5.8 million signatures. Here is a new way to involve the citizenry.

We also need to use these tools to turn the conversation about government to the positive and constructive. We spend too much time complaining about government and trying to catch the bastards red-handed. There are lots of red-handed bastards to catch. But some people in government do care and work hard. Until we expect the best of them, we will see only the worst. Let's think like engineers and identify problems and work toward collaborative solutions. Pollyannaish? Yes, but if we never move past complaining we'll never build anything new.

I'm not suggesting government should be crowdsourced. I don't want rule by the mob, even the smart mob. The internet requires filters, moderators, fact-checkers, and skeptics. So will the conversation that powers the country. That is the definition of a republic: representatives as filters. Those in power can use the internet to become better informed about our needs and desires and we can use it to speak and to contribute. The internet can transform the gift economy into the gift society.

The internet—which is so often accused of creating echo chambers where we hear only like minds—enables us to organize in new ways, around issues and not just party banners. People of any party or state, red or blue, can gather around the environment, taxes, education, health care, or crime as issues they want to tackle. This requires a new personal political openness: We need to say where we stand to find others who stand there. I'd like to see citizens use the web as personal political pages (PPPs) in which each of us may, if we choose, reveal our positions, opinions, and allegiances: the Facebook of democracy. I'd use a PPP to post my personal political statement online. In my case, I am a centrist Democrat; I voted for Hillary Clinton; I want to actively support movements to protect the First Amendment against Federal Communications Commission censorship; I believe we must support an aggressive national broadband policy; and I support universal health insurance. On my page, I would explain and discuss issues, linking to blog posts I've written or to others who speak effectively for my views. I already do this on the disclosures page of my blog because I try to practice transparency; my readers have a right to know where I stand on issues I write about so they can judge what I say accordingly.

On my PPP I should also be able to manage my relationship with politicians—a variation on the theme of Doc Searls' VRM or vendor relationship management. How about PRM: political relationship management? I want to say which candidates and organizations may approach me for my money or time. I'll invite opponents to try to convince me to change my mind: Give me your best shot. If someone convinces me, I'll change my public stance on the page. Personal political pages could

become a standard for disclosure and could be used to reveal in clear language the stands—as well as the conflicts and biases—of politicians and
journalists.

Let's imagine millions of these pages that can be searched and analyzed
to reveal a constant snapshot of the vox populi: Google as the polling
place that never closes, except now we control the questions and our opinions, not pollsters. This new public square makes politics and public opinion a constant process instead of an annual or quadrennial event. It is a
platform for organizing citizens. We can search Google for people who
agree on a topic and try to gather them around a page, petition, group,
politician, or organization.

When I toyed with this notion on my blog, one commenter, TV-industry
analyst Andrew Tyndall of the Tyndall Report, saw potential for reducing
the power of the left-right pigeonholes in which we're too often stuck.
Those pigeonholes, he said, make it

> so much more difficult to form coalitions with those at radically
> different parts of the ideological spectrum—with born-again Chris-
> tians who are leading activists on HIV/AIDS or Darfur genocide;
> with Wall Street free traders who want to liberalize immigration
> with Mexico; with Cato Institute libertarians who want to legalize
> narcotics; with centrist Democrats like Jeff Jarvis who want univer-
> sal healthcare; with neoconservative ideologues working to re-
> place autocrats and theocrats with democrats in the Middle East;
> with non-partisan bureaucrats like Michael Bloomberg who want
> to switch transportation from cars to mass transit.
>
> Personal political pages allow each of us to escape from the
> conventional left-right authoritarian-libertarian divisions of the
> political parties and the opinion pollsters. They allow us to align
> ourselves on each issue discretely, forming ad hoc, opportunistic
> coalitions not binding ones.

The moment Facebook was translated into Spanish (with the help of its
community), it was used to organize campaigns in Colombia against
FARC (Revolutionary Armed Forces of Colombia) guerillas. Facebook
was used to build a youth army for Barack Obama's run for the White

House. Facebook's Causes is used to help the public gather support for issues. The internet and Wikipedia are used to inform the electorate. Meetup is used to help organize voters. These are tools that can help us collaborate and manage our government. Google and company aren't taking over Washington. They're helping us take over.

Exceptions

PR and lawyers
God and Apple

PR and lawyers: Hopeless

When I suggested on my blog that there were three industries immune from rehabilitation through Googlethink, my readers disagreed about one—insurance, which spawned an earlier chapter. But nobody disagreed about PR and law. I won't turn this into a joke about flacks and lawyers—there are plenty of those already (go to Google, search for "lawyer jokes," and enjoy). Instead, I'll use this opportunity to examine a few of the key tenets and prerequisites of Googlification through the exceptions that prove the rules.

The problem for public relations people and lawyers is that they have clients. They must represent a position, right or wrong. As they are paid to do that, the motives behind anything they say are necessarily suspect. They cannot be transparent, for that might hurt their clients. They cannot be consistent, for they may represent a client with one stance today and the opposite tomorrow, and we'll never know what they truly think. In a medium that treasures facts and data, they cannot always let facts win; they must spin facts to craft victory. They must negotiate to the death, which makes them bad at collaboration. It's not their job to help anybody but their clients. They are middlemen. They won't admit to making mistakes well; clients don't pay for mistakes.

Having said that these folks can't be reformed according to Google's ways is not to say that they can't use the tools we've reviewed to their own benefit. Some already do. Many lawyers blog (see a selection at Blawg. com). Like venture capitalists, they find value in talking about their specialties, giving advice, attracting business, branding themselves, and sometimes lobbying for a point of view. Some can be counted on to cover legal stories with valuable experience, background, and perspective. Lawyers are a smart bunch who—surprise!—can write in English instead of legalese. Still, when a law blogger advises me to check my made-in-China tires for problems, I'm also aware that he's on the prowl for class-action clients. Law is business.

Some lawyers have taken advantage of online networking capabilities to create virtual law firms, eliminating the cost of offices and reducing the overhead of office staff. According to the blog Lawdragon, Virtual Law Partners uses these savings to give its partners 85 percent of billing revenue vs. the usual 30–40 percent. Virtual PR and consulting firms also operate loosely, bringing in members of their networks as needed for clients and communicating and collaborating without offices.

PR people are trying to use the tools of web 2.0, Google, search, and social media to update their practices. Many of them blog—see, for example, Richard Edelman, head of the eponymous PR firm, and his web 2.0 guy, Steve Rubel, who blogs, Twitters, and joins in any new digital fad that comes round the corner so he can educate clients about them. PR people use these tools to keep track of what is being said about their clients and to join in those conversations. They have also been burned. In 2006, two bloggers wrote about their cross-country RV tour of Wal-Marts, where they met no end of allegedly happy employees. Revealed to have been arranged by Edelman and paid for by the front organization Working Families for Wal-Mart, the tour turned out to be an old-fashioned PR stunt updated only with the use of blogs. Edelman fell on his sword in a blog post: "I want to acknowledge our error in failing to be transparent about the identity of the two bloggers from the outset. This is 100% our responsibility and our error; not the client's." Case in point: PR people are not, and likely cannot be, transparent. They have clients.

But it should be the job of PR advisers to convince clients that it is in their interest to be transparent and honest now that obfuscations and lies can be exposed so easily online. That is PR turned upside-down: Rather than representing and spinning the client to the world, they remind the client that the world is watching. They can also help companies fulfill their new role in the ecology of information online. We expect companies to have sites, to share information, to be factual if not fully transparent.

Openness is the best PR you can have. Still, because they only advise, PR people aren't often in a position to change how a company is managed.

I'm sure lawyers and PR people—like real-estate agents—will be glad to tell me where I'm wrong and I welcome that discussion on my blog: Let's have at it, and if there are ways to Googlify these trades, then congratulations. In the meantime, both fields need to watch out, for the tools of Google and the internet enable others to disintermediate, undercut, and expose them.

The law and its execution are aided by obfuscation. The internet can fix that. A small number of volunteers could, Wikipedia-like, publish simple, clear, and free explanations of laws and legal documents online. All it takes is one generous lawyer—not an oxymoron—to ruin the game for a thousand of them. I've seen a few such sites. They're not very good yet—none worth recommending—but they're a start.

Another trend that helps both lawyers and clients is the movement to open up laws and case law online, making them searchable and free. It is a scandal that the work of our own legislatures and courts is often hidden behind private pay walls. Westlaw and Lexis, the so-called Wexis duopoly, have turned our laws into their $6.5 billion industry. They add value by organizing the information, but others are now undercutting them. Forbes told the story of Fastcase, a start-up that uses algorithms instead of editors to index cases so it can reduce costs and lower fees to lawyers. Better yet, public.resource.org is fighting to get laws and regulations online for free. Patents are online now, and Google has made them searchable (go to google.com/patents and, for entertainment, look up pooper scooper—aka "Apparatus for the sanitary gathering and retention of animal waste for disposal" or "perpetual motion machine" or Google itself). Laws, regulations, and government documents are prime meat for Google's disintermediation.

Sometimes lawyers are employed merely to intimidate—but now the internet's power to gather flash mobs enables those targeted by attorneys to return the intimidation. I've seen many cases of bloggers pleading openly for help against big organizations that are threatening or suing them. They received offers of pro bono representation from lawyers, often thanks to the Media Bloggers Association. The intimidators then received floods of bad PR. The internet doesn't defang lawyers, but it can dull their teeth or bite them back.

I would like to see an open marketplace of legal representation—present your problem and take bids from lawyers who have handled similar cases, with data on their success rates. Legal representation can also be open-sourced. People who've been in cases can offer free advice and aid to others: Here's how I dealt with my landlord and here are all the documents I used; feel free to copy and adapt them.

The goal is to free the law—*our* law—from the private stranglehold of the legal priesthood. Between putting laws and cases online and making them searchable, creating simplified legal documents anyone can use, holding weapons to fight legal intimidation, and creating a more transparent marketplace, we would not replace the legal profession with all its faults but we could create checks on its power.

Even the Supreme Court could benefit from a little Googlification. After the Court's esteemed justices made two mistakes in two decisions one day in 2008—one in a case involving the death penalty and child rape, the other involving energy regulation—they were corrected by bloggers who would have been happy to do so *before* the decisions, if only they'd been given the chance. I wouldn't hold my breath.

What other industries are immune from Googlethink? VC Fred Wilson said construction, because it's so laden with atoms. Yes, but architecture is opening up—I've seen more than one effort to open-source both the creation and use of designs. We can also share ways to fix up our homes. Waste disposal? Atoms again, but I'll bet that we, the customers, will start using online soapboxes to gang up on manufacturers and force them to reduce their obscene packaging. Furniture? There's a blog called Ikeahacker that enables fans to share ideas for modifying the slavishly standardized Swedish products. Mining? The book *Wikinomics* delights in telling the story of a mining company that opened up its geologic data to enable the public to help it find deposits and to get a share of the wealth that resulted. Pornographers? Of course, they have been the pioneers in most every innovation in online media and the industry benefited from each move—until amateur porn came along on PornTube (the not-safe-for-work YouTube) to undercut the business benefits of scarcity. The military? Actually, it was among the earliest users of blogs and wikis because it wants troops to share their experience and what they know. Terrorists? Unfortunately, they have made all-too-effective use of the internet and SEO to spread poison and create networks. No, few are immune from Google's impact.

God and Apple: Beyond Google?

OK, then, what about God? Is he immune from Googlethink? Churches have used the internet to spread their word and create virtual congregations that meet online or through Meetup. There are religious versions of many of the internet's big sites—such as GodTube, holier than YouTube—and religious groups have made clever use of others: God is big on MySpace and Facebook. Bible and Koran verses are searchable not only on the web but even on the iPhone. It's hard to imagine God endorsing a wiki version of the Bible—but then, wasn't the Talmud the world's first wiki? There are even web 2.0 religious movements. Open-Source Judaism—inspired by Douglas Rushkoff's 2003 book *Nothing Sacred: The Truth About Judaism*—has created the Open-Source Haggadah (a prayer book). God is not immune from the power and influence of Google.

Is there any entity that is untouched? Is there an anti-Google, one institution that has become successful by violating the rules in this book? I could think of one: Apple.

Consider: Apple flouts Jarvis' First Law. Hand over control to the customer? You must be joking. Steve Jobs controls all—and we want him to. It is thanks to his brilliant and single-minded vision and grumpy passion for perfection that his products work so well. Microsoft's products, by contrast, operate as if they were designed by warring committees. Google's products, though far more functional than Microsoft's and built with considerable input from users, appear to have been designed by a computer (I await the aesthetic algorithm).

Apple is the opposite of collaborative. It's not that it doesn't care what we think. After a product comes out, Apple has learned to fix its mistakes—quietly. The first iPhone's headphone jack was recessed in the case to make it look prettier, but that also made it incompatible with all plugs but Apple's own. In the next iPhone, the problem was fixed. Make mistakes well? Apple makes them quietly. Apple has apologized—most recently for its botched MobileMe launch—but mea culpas are rare.

Apple is a cult company and its customers are its best marketers—that much is Googley. Apple customers have made commercials for its products, they love them so. But Apple still spends a fortune in advertising, imbuing the brand with more cool because its commercials are as well-designed and well-executed as its products. Its most effective advertisement of all is Jobs' keynote lecture and demonstration at Apple conferences. The company could not be more one-way and less interactive.

Apple is the farthest thing from transparent. It has sued bloggers for ferreting out and revealing its secrets. Attacking its own fans was unbloggy and uncool, but Apple didn't care about the bad publicity. It's Apple.

Apple abhors openness. That's another reason its products work so well, because it controls what can run on them, how it runs, and how it makes money. When the iPhone came out, there were many complaints from open-minded geeks about not being able to install their own programs. Then with the next iPhone, Apple created a closed app store with lots of choices. The complainers kept themselves busy trying out new toys, and many said it was a pleasure to see applications that had been screened for quality, unlike the software fleamarket that Facebook and MySpace had become.

Apple's closed way of doing business is one of its advantages. While the rest of the online world was merrily destroying the music business with openness, Apple created the secure means for fans to buy billions of songs legally and happily.

Apple does, however, support open-source software, bragging on its site that it contributes to dozens of pools of code. That is a good business decision. Apple based its operating system on Unix rather than trying to make one itself; it's cleaner, far more reliable, and simpler than Windows. Apple's not stupid.

Apple does not think distributed. It makes us come to worship at its altar.

Apple does not manage abundance. It creates scarcity. Witness the fanatics who camped out overnight to get each version of the iPhone. According to blog reports, the company cut off sales of the phones on the first day with devices still in stock so there would be lines again the second day. Apple makes its own mobs.

Atoms? Apple has no problem with them. iTunes drives customers to buy more Apple hardware.

Free as a business model? The gift economy? Apple is not generous. It charges a premium for its quality.

Apple follows just a few Google rules. Lord knows, it innovates. And nobody's better at simplifying tasks and design.

How does Apple do it? How does it get away with operating this way even as every other company and industry is forced to redefine itself? It's just that good. Its vision is that strong and its products even better. I left Apple once, in the 1990s, before Steve Jobs returned to the company, when I suffered through a string of bad laptops. But when I'd had it with Dell, I returned to Apple and now everyone in my family has a Mac (plus one new Dell); we have three iPhones; we have lots of iPods; I lobbied successfully to make Macs the standard in the journalism school where I teach. I'm a believer, a glassy-eyed cultist. But I didn't write this book about Apple because I believe it is the grand exception. Frank Sinatra was allowed to violate every rule about phrasing because he was Sinatra. Apple can violate the rules of business in the next millennium because it is Apple (and more important, because Jobs is Jobs).

So then Apple is the ultimate unGoogle. Right?

Not so fast. When I put that notion to Rishad Tobaccowala, he disagreed and said that Apple and Google, at their cores, are quite alike.

"They have a very good idea of what people want," he said. Jobs' "taste engine" makes sure of that. Both companies create platforms that others can build upon—whether they are start-ups making iPod cases and iPhone apps or entertainment companies finding new strategies and networks for distribution in iTunes.

Apple, like Google, also knows how to attract, retain, and energize talent. "Apple people believe they are even better than Google people," he said. "They're cooler."

Apple's products, like Google's, are designed simply, but Tobaccowala said Apple does Google one better: "They define beauty as sex," he said.

Apple understands the power of networks. Its successful products are all about connecting. Apple, like Google, keeps its focus unrelentingly on the user, the customer—us—and not on itself and its industry. And I'll add that, of course, both companies make the best products. They are fanatical about quality.

But Tobaccowala said that what makes these two companies most alike is that—like any great brand—they answer one strong desire: "People want to be like God." Google search grants omniscience and Google Earth, with its heavenly perch, gives us God's worldview. Apple packages the world inside objects of Zen beauty. Both, Tobaccowala said, "give me Godlike power." WWGD? indeed.

Generation G

Google is changing our societies, our lives, our relationships, our world-views, probably even our brains in ways we can only begin to calculate.

Start with our relationships. I believe young people today—Generation Google—will have an evolving understanding and experience of friendship as the internet will not let them lose touch with the people in their lives. Google will keep them connected. Admit it: You've searched for old girlfriends and boyfriends on Google (and wondered whether they've Googled you). Your ability to find those old, familiar faces likely drops in inverse proportion to age: The older you are, the harder it is to find old friends online. I went to Google—purely as an academic and technical exercise, understand—and searched for old girlfriends. I found my college girlfriend, now a philosophy professor. I couldn't find my high-school sweetheart as she had left no visible Google tracks. But she later found me because, with my blog, I had left as many tracks as a herd of buffalo in snow. We live on opposite coasts now but when I was in her city on business, we got together and filled each other in on the last—gulp—30-odd years. We never would have had that chance to catch up and come to account without Google. Thank you, Google.

That won't be the experience of young people today. Thanks to our connection machine, they will stay linked, likely for the rest of their lives. With their blogs, MySpace pages, Flickr photos, YouTube videos, Seesmic conversations, Twitter feeds, and all the means for sharing their lives yet to be invented, they will leave lifelong Google tracks that will make it easier to find them. Alloy, a marketing firm, reported in 2007 that 96 percent of U.S. teens and tweens used social networks—they are essentially universal—and so even if one tie is severed, young people will still be linked to friends of friends via another, never more than a degree or two apart.

I believe this lasting connectedness can improve the nature of friendship and how we treat each other. It will no longer be easy to escape our pasts, to act like cads and run away. More threads will tie more of us together longer than in any time since the bygone days when we lived all our lives in small towns. Today, our circles of friends grow only larger.

Does this abundance of friendship make each relationship shallower? I don't think so. Friendship finds its natural water level—we know our capacity for relationships and stick closest to those we like best. The so-called Dunbar number says we are wired to pay attention to about 150 relationships. I think that could grow with relationships of various kinds that are easier to maintain online. But remember the key insight that made Facebook such a success: It brought real names and real relationships to the internet. It's about good friends.

Won't our embarrassments also live on? Our missteps, youthful mistakes, and indiscretions will be more public and permanent, haunting us for the rest of our lives because the world, thanks to Google, has a better memory. True. But here the doctrine of mutually assured humiliation enters to shield us. We will all have our causes to cringe. The tarnished flipside of the golden rule becomes: I'll spare you your shame if you spare me mine. Or to put it more eloquently, I once again quote author David Weinberger, who said at a conference (according to the Twitter feed of blogger Lisa Williams, who was there): "An age of transparency must be an age of forgiveness." Our new publicness may make us more empathetic and ultimately forgiving of each others' and even of public figures' faults and foibles. We see that already. Barack Obama said he inhaled and no one gasped. Who are we to throw stones when Google moves us all into glass towns? In Googley terms: Life is a beta.

But still, I hear, hasn't life become too public? What has become of privacy? "Nothing you do ever goes away and nothing you do ever escapes notice," Vint Cerf, one of the fathers of the internet and most recently a Google executive, told an audience in Seattle. Then he added—please note, with irony—"There isn't any privacy, get over it." He's right. I say privacy is one of the most overused fear words of the age. Privacy is not the issue. Control is. We need control of our personal information, whether it is made public and to whom, and how it is used. That is our right, at least for matters outside the public sphere.

The ethics and expectations of privacy have changed radically in Generation G. People my age and older fret at all the information young people make public about themselves. I try to explain that this sharing of personal information is a social act. It forms the basis of the connections Google makes possible. When we reveal something of ourselves publicly, we have tagged ourselves in such a way that we can be searched and found under that description. As I said in the chapter on health, I now can be found in a search for my heart condition, afib. That is how others came to me and how we shared information. Publicness brings me personal benefits that outweigh the risks.

Publicness also brings us collective benefits, as should be made clear by now from the aggregated wisdom Google gathers and shares back with us thanks to our public actions: our searches, clicks, links, and creations. Publicness is a community asset. The crowd owns the wisdom of the crowd and to withhold information from that collective knowledge—a link, a restaurant rating, a bit of advice—may be a new definition of anti-social or at least selfish behavior.

For all these reasons and one more powerful than any of them—ego—we will continue to reveal more of ourselves online. We will want to speak and to be discovered. Our online shadows become our identities. To stand out from our crowd, we need distinct identities. I'll bet we'll soon see parents giving children unique names so they can stand alone in Google searches. Wired editor Chris Anderson linked to an early indication of the trend: Laura Wattenberg, author of *The Baby Name Wizard*, reported that in the 1950s, a quarter of all children got one of the top 10 baby names; more recently that has fallen to a tenth. I was about to predict that some-day soon, parents would check to assure the .com domain for a name is available before giving the moniker to a baby. Then I searched on Google and, sure enough, the Associated Press reported in 2007 that it's already happening: "In fact, before naming his child, Mark Pankow checked to make sure 'BennettPankow.com' hadn't already been claimed. 'One of the criteria was, if we liked the name, the domain had to be available,' Pankow said." At last check, young Bennett wasn't blogging, but his digital destiny is set.

More than names, identity will be about accomplishments and creations, things you are known for that narrow your Google search. I am the blogger Jeff Jarvis who writes about Google and media, not Jeff Jarvis the jazz trumpeter, Jeff Jarvis who ran Segway tours in Thailand (drat—I think I'd like to be him), Jeff Jarvis who heads a mobile field service software provider (whatever that is), and certainly not Jeff Jarvis the high school athlete (sadly, I'm too old and too clumsy). I am the No. 1 Jeff Jarvis. In Google wars, it's every Jeff for himself.

This brings us to another argument against public identity: It turns us into egotistical exhibitionists. We share everything, down to the most intimate and mundane. Who cares what I had for breakfast? Why share it? London blogger Leisa Reichelt found that this "ambient intimacy"— reporting small signposts of life, sharing what we're doing, who we're with, when we get a new haircut or a new car—allows us to "keep in touch with people with a level of regularity and intimacy that you wouldn't usually have access to, because time and space conspire to make it impossible." Ambient intimacy is good for friendship. "It helps us get to know people who would otherwise be just acquaintances. It makes us feel closer to people we care for but in whose lives we're not able to participate as closely as we'd like." And on a practical level, Reichelt said, "It also saves a lot of time when you finally do get to catch up with these people in real life!"

The internet and Google are causing no end of small behavioral changes whose impact is, again, difficult to weigh quite yet. Some may be short-lived fads; others may have a long-term influence on societal norms. Here are a few:

- Ever since I started working with computers, I've found it terribly seductive that there is always a solution to a problem involving machines and software. You just have to find it. If only life were so symmetrical and complete. I fear young people today could become more disappointed with the harsh, illogical, and incomplete reality of life than my generation was. Then again, we were flower children.

- I wonder, too, whether Google's slavish devotion to data, its belief that numbers tell truth, could have us miss the qualitative, counterintuitive, human view of life: the eureka moments that come from the illogical. Would we still discover the accidental gift of bread mold, penicillin?

- Then again, perhaps all this will hone our analytic skills. Employees at Google are not permitted to rely on intuition, hunches, wishes, beliefs, and the way things have always been—easy answers and accepted wisdom. Perhaps our employees, bosses, politicians, and educators would better serve us if they were held to such an empirical standard.

- I would be delighted if education put less emphasis on rote memorization of that which we can easily look up, but I wonder whether Google's instant access to every imaginable fact will atrophy our memory cells. Or perhaps that's just my fear of age.

In a 2008 article in The Atlantic, internet curmudgeon Nicholas Carr, a sometimes sparring partner of mine in the blogosphere, fretted about these changes in our habits, brains, and society in an article entitled, "Is Google Making Us Stupid?" He confessed to reading less and differently—as I have. "The kind of deep reading that a sequence of printed pages promotes is valuable not just for the knowledge we acquire from the author's words but for the intellectual vibrations those words set off within our own minds," Carr argued. "In the quiet spaces opened up by the sustained, undistracted reading of a book, or by any other act of contemplation, for that matter, we make our own associations, draw our own inferences and analogies, foster our own ideas. Deep reading, as Maryanne Wolf argues, is indistinguishable from deep thinking."

Google CEO Eric Schmidt's defense against Carr: "I observe that we're smarter than ever." Carr might accuse me of triumphalist reflex—it wouldn't be the first time—but I say that deep interaction, too, can yield deeper thinking. Because I write in short blog bursts instead of long essays, it might appear that my thoughts are quicker and shallower—you're free to conclude that. But my ideas may span many posts and take form and shape over weeks and even months, with input, challenge, and argument from many of my blog readers and commenters. Under that pressure, I also drop ideas that don't work. For me, the blog is a new and efficient means of both collaboration and peer review. It molded a great many of the ideas in this book. So though I do fret about the unread books on my shelves and the virgin New Yorker magazines on my desk—as well as a constant stock of unread tabs in my browser—I also know that I learn volumes online every day. Is what I do now better or worse? I'm not sure that judgment is meaningful. I learn differently, discuss differently, see differently, think differently. Thinking differently is the key product and skill of the Google age.

It has been said that young people today may take on new behavioral norms and mores and political outlooks from games and social software—and I don't mean sex and violence, but subtler worldviews. "Social software is political science in executable form," NYU professor Clay Shirky said in one essay. "Social norms in game worlds have the effect of governance," he said in another. Stanford law professor Lawrence Lessig famously declared that code is law: "This code, or architecture, sets the terms on which life in cyberspace is experienced. It determines how easy it is to protect privacy, or how easy it is to censor speech. It determines whether access to information is general or whether information is zoned. It affects who sees what, or what is monitored." He said code "implements values, or not. It enables freedoms, or disables them."

So what ethics, values, mores, and models are implicit in Google and our use of it, and how might they affect Generation G? Once more the caveat: It's difficult to know. But we can speculate. I talked earlier in the book about the ethics I learned from blogs and bloggers: the ethics of the link, of transparency, and of the correction. What else flows from Google?

- I believe the aesthetic of simplicity we see online becomes an ethic of simplicity. Elegant code is spare and efficient. That norm of geek culture carried over to Google's home page and design, where powerful tasks look unassuming and easy—simplicity is complexity well done. Simplicity may carry over from web sites to products to culture and to our view of life.

- Google rewards—and more and more, we expect—openness. In our lives, openness takes the form of personal transparency—the bloggers' code that calls for revealing one's conflicts and prejudices. In business, companies built on proprietary secrets may not be trusted. The public will now expect them to operate in the open.

- I think we will see growing respect for the small and odd. The mass norm of keeping up with the Joneses now yields to prideful individuality because Google rewards uniqueness in the mass of niches—and because odd geeks are coming to rule the culture.

- But I fear ours could become a culture of complaint—and I would bear some personal responsibility for that, given my Dell battle. Online, complaint pays off, and after so many years of being subjugated to corporate control, it feels mighty good for us little guys to win. But online, any complaint also threatens to become a war. We, the people, have to learn that we have more power than we know, and we must learn to use it judiciously.

Out of all the new societal norms the internet fosters, my greatest hope is that future generations will enforce a doctrine of free speech with governments and institutions. The internet is the First Amendment brought to life. It abhors and subverts censorship—for whatever speech is tamped down in one place can and will arise somewhere else. That is the positive force of global communication. The danger in this globalism, however, is that our freedom could be reduced to a lowest common denominator of speech as dictated by the worst regime, whether that is through government repression, pressure groups that object to TV shows in America or cartoons in Denmark, or regressive libel laws (which some say are outmoded now that everyone has the means to respond). We must expect powerful forces such as Google to use their economic, cultural, and moral influence to pressure censors in China, Iran, and elsewhere to value and protect speech.

Whatever causes they take up, Generation G will be able to organize without organizations, as Shirky wrote in *Here Comes Everybody*. That ability to coalesce will have a profound destabilizing impact on institutions. We can organize bypassing governments, borders, political parties, companies, academic institutions, religious groups, and ethnic groups, inevitably reducing their power and hold on our lives. In an essay in Foreign Affairs in 2008, Richard Haass argued that the world structure is moving from bi- and unipolarity (i.e., the Cold War and its aftermath) to nonpolarity (i.e., no one's in charge). We now operate in an open marketplace of influence. Google makes it possible to broadcast our interests and find, organize, and act in concert with others. One need no longer control institutions to control agendas. Haass chronicles the dilution of governments. Bloggers Umair Haque and Fred Wilson have written about the fall of the firm, and earlier I examined the idea that networks are becoming more efficient than corporations. In my blog, I follow the crumbling of the fourth estate, the press. One could debate the stature and power of the first estate, the church. What's left? The internet is fueling the rise of the third estate—the rise of the people. That might bode anarchy except that the internet also brings the power to organize.

Our organization is ad hoc. We can find and take action with people of like interest, need, opinion, taste, background, and worldview anywhere in the world. I hope this could lead to a new growth in individual leadership: Online, you can accomplish what you want alone and you can gather a group to collaborate. Being out of power need not be an excuse or a bar from seeking power. That may encourage more involvement in communities and nations—witness the youth armies that gathered in Facebook around Barack Obama, a powerful lesson for a generation to have learned.

Early in its rise, I wondered whether the internet would be inherently liberal or conservative. Conventional wisdom says that broadcast TV, serving the masses, was the medium of the left, whereas talk radio and cable TV, serving large niches with the ability to hammer contrary messages, were the media of the right. What is the internet, then? At first, I thought it was libertarian as that was, disproportionately, the ethos of so many early political bloggers. It made sense: The internet champions and enables personal liberty. But as time went on, I learned that the internet is neither a monolith nor a medium. In industry and politics, it disaggregates elements and then enables free atoms to reaggregate into new molecules. It fragments the old and unifies the new. It obsoletes old orthodoxies and old definitions of left and right and provides the opportunity to make more nuanced expressions of our political worldview. It was then that I saw the internet not as left, right, or libertarian but as the connection machine that brings together any and all worldviews.

I pray that Google and the internet will change, spread, and strengthen democracy. Google's moral of universal empowerment is the sometimes-forgotten ideal of democracy. This revolution won't start at the top, in governments and institutions. As with everything Google touches, it will grow from the bottom, in communities of all sizes and descriptions, as more involvement leads to new ways to organize, manage, and govern. That is what we mean when we talk about power shifting to the edge, no longer centralized. Political movements need not start in Washington but can start in a thousand places linked online. When millions of people give $10 each to a campaign—instead of 10 people giving $1 million each—the power in a party shifts to the edge, some hope. That is what political strategist Joe Trippi argues in his book, *The Revolution Will Not Be Televised*. Generation G will have a different sense of membership, loyalty, patriotism, and power. They will belong to new nations: a nation of geeks, a nation of diabetics, a nation of artists. They may feel greater allegiances to these nations and less to their town or country.

Hear the Declaration of Independence of Cyberspace by John Perry Barlow, former Grateful Dead lyricist and a founder of the Electronic Frontier Foundation, from 1999: "Governments of the Industrial World, you weary giants of flesh and steel, I come from Cyberspace, the new home of Mind. On behalf of the future, I ask you of the past to leave us alone. You are not welcome among us. You have no sovereignty where we gather." Barlow warned that the old world's laws of property, identity, and movement "are all based on matter, and there is no matter here." He said the only law that all online cultures recognize is the golden rule. "We are creating a world where anyone, anywhere may express his or her beliefs, no matter how singular, without fear of being coerced into silence or conformity."

My generation, the children of the sixties, prided itself on nonconformity, but our nonconformity became conformist. I fear it was a fashion. Some worry that Generation G's nonconformity and individualism will be entitled rather than empowered, alone more than social, entertained more than educated. Any of that and worse could be true. But I have faith in this generation because, far earlier than their elders—my peers—today's young people have taken leadership, contributed to society and the economy, and created greatness: great technology, great companies, great thinking.

That is where we return at the end: creation. Looking at the internet, one must be struck by the will of the people to create. One survey I quoted earlier reported that most of us say we have a book in us. Another said, coincidentally, that most of young people think they have a business in them. We have surveyed our creation: We make tens of millions of blogs. We take hundreds of millions of Flickr photos. A few hundred thousand people write applications for Facebook. Every minute, 10 hours of video are uploaded to YouTube. People create T-shirt designs on Threadless, sneaker designs on Ryz, and things of all descriptions on Etsy. Kids make companies. And on and on.

The internet doesn't make us more creative. Instead, it enables what we create to be seen, heard, and used. It enables every creator to find a public, the public he or she merits. That takes creation out of the proprietary

hands of the supposed creative class. Internet curmudgeons argue that
Google and the internet bring society to ruin because they rob the creative class of its financial support and exclusivity: its pedestal. But internet
triumphalists, including me, argue that the internet opens up creativity
past one-size-fits-all, mass measurements and priestly definitions of quality and lets us not only find what we like but also find people who like
what we do. The internet kills the mass, once and for all. With that
comes the death of mass economics and mass media. I don't lament their
passing.

There will still be a creative class, but it's role and relationship with the
public may change, acting not just as creators but also as examples, educators, and inspirations for others—the flint of creativity. That is what Paulo
Coelho became when he asked his readers to make a movie of his book.

The curmudgeons also argue that this level playing field is flooded with
crap: a loss of taste and discrimination. I argue instead that only the playing field is flat. To stand out, one must rise on worth—as defined by the
public rather than the priests—and the reward is attention. That is our
culture of links and search. It is a meritocracy, only now there are many
definitions of merit and each must be earned.

We have believed—I have been taught—that there are two scarcities in
culture: talent and attention. There are only so many people with talent
and we give their talent only so much attention—not enough of either.
But just as the economy is shifting from scarcity to abundance, so is the
culture. There is an abundance of talent and a limitless will to create, but
they have been tamped down by an educational system that insists on
sameness, starved by a mass economic system that rewarded only a few
giants, and discouraged by a critical system that anointed a closed creative
class. These enemies of mass creativity turned abundance into scarcity.
Google and the internet reversed that flow. Now talent of many descriptions and levels can express itself and grow. We want to create and we
want to be generous with our creations. We will get the attention we deserve. That means crap will be ignored. It just depends on your definition
of crap.

When we talk about the Google age we are talking about a new society.
The rules explored in this book—Google's rules—are the rules of that
society, built on connections, links, transparency, openness, publicness,
listening, trust, wisdom, generosity, efficiency, markets, niches, platforms,

networks, speed, and abundance. This new generation and its new world-view will change how we see and interact with the world and how business, government, and institutions interact with us. It is only just beginning. I wish I knew how that change will turn out. But I'm thrilled to be here today with you to witness its birth.

Afterword

The best part of writing *What Would Google Do?* came after it was published, when people from a surprising range of sectors told me how they had tested the rules described in this book in their own endeavors.

I spoke at a convention of truck-stop owners who realized that their way stations could act as nodes to build networks among drivers who have information to share with each other. *Join a network. Be a platform. Think distributed.*

At the other end of the demographic spectrum, I heard from executives at two of the largest luxury-goods companies in the world, who saw value in opening up their exclusive design processes in order to connect with new tastemakers and new talent and become curators of quality and luxury. *Elegant organization.*

From yet another category, I heard foundations speculate on how different their work would be if they opened up their structures to identify new needs, new grantees to meet those needs, new ways to measure their success, and new ways to leverage their assets by encouraging others to help in their work. *Join the open-source, gift economy.*

At a meeting of librarians, we faced their worst-case scenario—library closings—and then catalogued the value librarians continue to provide when information and search functions are digital but human expertise and guidance aren't. *Atoms are a drag.*

A group of postal executives wondered what Google would do if it ran the post office. One official speculated that it would give every American a computer and printer, replacing mail and slashing costs. This discussion led to the PostalVision 2020 Conference in Washington, where I pushed the industry not to try to fix the postal service one cutback at a time but instead bravely consider what the market could do on its own. *Beware the cash cow in the coal mine. Do what you do best and link to the rest. Get out of the way.*

At the height of the financial crisis, I moderated a session at Davos in which entrepreneurs discussed how to fix the broken banking industry. They imagined a bank that would be open with all its data, from investments to salaries. *Be honest. Be transparent. Don't be evil.*

Lufthansa ran a brainstorming session with a score of social-media practitioners at the DLD Conference in Munich, to find out how even an airline could be Googley. The bottom line: If airlines share information with customers (why *is* the plane late?), customers will be more willing to share information with airlines, provided that airlines make good use of it (for example, assigning me the exact seat I like best). *There is an inverse relationship between control and trust.*

Best Buy's tweeting chief marketing officer, Barry Judge (@BestBuy-CMO), invited me to the company's headquarters to try out some of the ideas there. I learned more from them than they did from me as I observed how a smart company moves from simply selling things in boxes to providing service and expertise. Best Buy opened up its infrastructure to allow others to build onto it. Three thousand sales people answer customers' questions through a single Twitter account (@Twelpforce), forming a human search engine. Best Buy is also becoming a media company as it sells promotional opportunities in its stores. *Decide what business you're in.*

Sales guru and author Jeffrey Gitomer introduced me to his staff to help them decide how they could be more Googley. I suggested they start by gathering the best sales tips from their own readers, who are out there selling and succeeding every day. Gitomer's own blogs and tweets have inspired his latest book, *Social BOOM!*, about this new way of doing business. *Trust the people. Your customers are your ad agency.*

In my next book, *Public Parts*, I tell the story of a very Googley car company, Local Motors, which designs cars openly. *Collaborate.* I report on visionaries who are rethinking retail from the ground up, now that

Google and the internet are make pricing transparent. *Google commodifies everything. Welcome to the Google economy.*

I have even learned of church pastors who aspire to be Googley by using the web as a tool to leave their brick walls behind and go to where their parishioners live. Church magazine suggests a "move from giving answers to asking questions." *Listen. Trust the people. Everybody needs Googlejuice.*

These church folks did not fall for the joke in the title of this book. Google isn't God, and the laws here are not immutable. "We don't consider Jarvis' rules to be sacred or unchanging," Leisa Anslinger and Daniel S. Mulhall wrote in the magazine, "but they do provide a valuable tool to help us rethink how we are to be a church in the twenty-first century."

It is with some considerable relief that when I reread *What Would Google Do?*, I find that its gospel still stands. But as I said at the beginning, this is less a book about Google than it is about the changes overtaking our world. Those changes only prove to be more disruptive—and more important to understand—by the day.

Print journalism, retail, automobile manufacturing, banking, telecommunications, and real estate have all suffered greater and faster upheaval than I would have predicted when writing *What Would Google Do?* I do not believe this disruption came from the financial crisis, which was entering its full flourish as I finished the book. I don't believe the disruption can be quelled with time or infusions of cash that can, say, create a few more jobs by fixing a few bridges. No, we are at the cusp of a profound restructuring of the economy—which I write about in this book—and of society—which I write about in my next book, *Public Parts*.

We see old corporations, industries, and institutions trying to ignore, resist, then fight the change overcoming them—futilely. I see that behavior at closest range in my own industry, the news business, where newspapers are finally recognizing that they will shrink in print. But they still fail to understand, as Google does, that a media economy built on scarcity is being replaced by one built on abundance. Editors don't control information anymore, publishers have lost their pricing power, and none of them knows what to do about it. That is why I have chosen to turn my attention to teaching entrepreneurial journalism, because it is apparent that innovation will come mostly from new perspectives rather than old powers.

Drive on any commercial highway in America and you see the slow collapse of retail: empty boxes that used to house Circuit City stores, supermarkets, and bookstores. Amazon is partly to blame, as it exploited new efficiencies in consolidating and reducing inventory. It used its scale to make delivery affordable. It reduced risk. Data is Amazon's greatest strength; it knows what we buy so it can sell us what we want. Data is also a key catalyst in retail's decline: Our ability to search Google for any product and find the lowest price has robbed stores of their ability to arbitrage opacity—that is, to profit by our ignorance of prices elsewhere. Google is constantly finding new ways to make markets even more transparent. Now you can pick up your smartphone and with Google Goggles take a picture of an item to search for better prices. Retail's decline will continue.

The automobile industry collapsed under the weight of its own inefficiency, high labor and benefit costs, new competition, lack of innovation, poor quality, and mismanagement. But the car companies, like the banks and financial institutions, were judged too big to fail, and so they were bailed out with tax dollars. Should we have intervened in their fall?

I asked Eric Schmidt this question at the Aspen Ideas Festival in 2009. From the audience, I proposed the idea that we were passing from an industrial age built on mass production, distribution, marketing, and media into an economy built on knowledge and abundance, and that companies such as Google should be encouraged to build new versions of our dying industries. "Am I going too far?" I asked. "I think so," he said—and then, having read this book, joked, "of course, you're good at it. Your book was about taking some ideas and talking about them in a global context."

Then Schmidt continued: "While I'd like what you said to be true, it's not today yet true. I'd like us to make it true. The reason is that almost all the money, almost all the people, almost all the capital is not going to where you described. It's going into traditional businesses." He said that entrepreneurs trying to disrupt stagnant industries hit regulatory and other barriers. "The incumbents, typically large companies, working with regulators, have ended up making a cozy structure for themselves. And when the truly discontinuous idea comes along, it's not in anyone's interest to take it on. . . . Governments are not particularly good at dealing with change." Is Google "Google" only because it is wholly new? I asked.

"I would argue that Google is as successful as it is primarily because of the openness of the internet," he replied.

What makes Google successful is, of course, the question that underlies this book. So now I must ask again whether Google itself can still be held up as the exemplar of a company that has figured out the new age. All in all, I think it can be. It continues to dominate its markets, innovate as it enters new businesses, and grow (zooming from $21.8 billion in revenue in 2008 to more than $34 billion, annualized from the first quarter of 2011).

But as I said earlier, Google's preeminence is by no means assured. No company's is. Since writing *WWGD?*, Myspace has all but fallen into a rabbit hole and Digg has not met the high expectations I had for it here (Kevin Rose has left the company). Skype has been bought and sold and bought again, its value rising and falling and rising again along the way. There are no guarantees, no safe bets.

Google has made missteps. It created its version of a social messaging service, its Twitter, called Buzz, and stepped into a privacy hornets' nest when Gmail users found their email contacts used as if they were all friends. Google made the mistake of testing the service only with its own employees and missing the obvious: that Googlers are not typical. Google thus violated one of the key rules in this book: Buzz was the first major product Google did not release as a beta, yet it was the product that most needed the collaboration and learning such a period of open development would have allowed. (I'll refrain from suggesting that more Google executives should have read this book!)

Google fumbled again when it decided to digitize the world's books. A furor about copyright ensued, and negotiations over a settlement continued in and out of court. I believe Google's mistake was in taking possession of content. It was meant only to organize the world's content, and when Google got too close to controlling any of it, conflict, mistrust, and fear were inevitable—especially given that Google had become so big and powerful. I don't believe Google's motives were bad. In this book, I complain that print books are in many ways inferior to knowledge in digital form. The world wasn't digitizing its books, and so Google stepped in to take on the task. Even as Google was portrayed as a Godzilla, I think it saw itself more as Snuffleupagus: big, clumsy, cuddly, and helpful, but unaware of its size.

In *WWGD?*, I said that Google's greatest challenge was its size. Even so, I underestimated how being big can prove to be both an advantage and a disadvantage. Schmidt often says his company's greatest challenge is its growth. Once a company is big, it becomes harder for that company to keep growing. As it grows, a company passes an invisible line, over which critics begin to complain that it has become too big. If our banks were too big to fail, Google, in the eyes of some, became too big to be allowed to succeed. That is one reason why Google has attracted the glare of regulators across the United States and Europe over issues of advertising, privacy practices, and even pictures of public places on Google Street View.

A gigantic company—especially one that explodes so rapidly—is obviously difficult to manage. That ungainliness is likely to blame for Google's greater rate of failure with new products, including Buzz; the live, collaborative platform called Wave; the Foursquare predecessor (from Foursquare's own founder), Dodgeball; a service meant to allow readers to add comments to any web page called Sidewiki; Google Video (finally folded years after the company acquired YouTube); and Google TV, which was universally panned for being too difficult to use. A company that makes innovation a key strength with its 20 percent rule is bound to fail at many experiments. But the rate of failure seemed to increase as both the right hand and left hand spent too much time groping to find each other and a cohesive strategy.

In 2011, Google's management triumvirate—founders Larry Page and Sergey Brin and CEO Eric Schmidt—conceded that change was needed at the top. Schmidt said his task of giving the founders "adult supervision" was no longer needed, so he kicked himself upstairs to become executive chairman and handed the keys to the (solar-powered, computerized, self-driving) car to Page. Analysts hope that having a single head of the company rather than a trilogy will yield more decisive strategy and execution. We shall see.

Since writing *WWGD?*, I have said that there are three wars Google has not won: for the local, live, and social web. The social battle was all but conceded to Facebook and the live fight to Twitter, although Google+ is a credible combatant in each arena. But local is just now being fought, and the battleground is mobile. Once we are all connected everywhere and all the time, who's to know whether we connect from devices at home or out on the street? Mobile is coming to mean local: Google and Apple are competing to be with us everywhere we go, depending on which phones we choose.

In this area, Google has made an impressive transformation—known as a pivot in Silicon Valley—from a search and advertising company to a mobile company, with the release of its open-source Android operating system for phones, tablets, and untold devices yet to be imagined. Google followed one of the rules in this book—asking what business it is really in—and moved beyond search engine as its essence. It recognized the value of gathering more information about its customers by serving us all the time, wherever we go.

In the mobile category, the contrast with Apple's methods came into starker relief: Apple's iPhone and iPad succeed because they are so tightly controlled to ensure quality. Google's strategy is to use openness to be on as many devices as possible, attracting as many users as possible, enticing as many developers to create as many apps as they can to attract more customers, and so on.

It is in mobile where Google's battles with Facebook, Foursquare, and other digital companies will also take shape as each tries to learn more about us—where we are, what we're doing, what we're looking for, who we're doing it with, what we say.... Each company wants to use these bits of data about us so it can best target content, services, and advertising to our needs and desires. And it is in local markets that Google hopes to find its next pot of gold as it competes with these social companies as well as with coupon seller Groupon, local content company Patch (part of AOL), and also the (gasping) newspapers, all of which are trying to figure out how to scale the sales of digital adverting to countless local merchants.

Remember, I chose not to write Google's story here. Others have done that, and since this book was published, two more have joined the ranks of Google's Boswells: Ken Auletta (*Googled: The End of the World As We Know It*) and Steven Levy (*In The Plex: How Google Thinks, Works, and Shapes Our Lives*). I chose instead to survey Google from a distance, to reverse-engineer its success to define the rules that could make anyone succeed in this era of volcanic change and explosive opportunity. That confused a few critics, who thought I was suggesting that we should all *be* Google. No, I was suggesting that we all could learn lessons from Google's worldview and its methods, particularly its ability to see and understand the change occurring around us in a new light and thus to find and exploit new opportunities.

And so it was only after writing the book that I visited Google. I gave a book talk in its Mountain View headquarters, the Googleplex, which was rather unnerving. Here I was in a room filled with proven geniuses who could tell me exactly how wrong I was. Yet they didn't, and I hope that wasn't because they were just being polite. I left impressed mostly at how Google defaults to intelligence. Most of us who've labored like Dilberts inside big companies begin to assume that the machine we're in tends toward the stupid. But at Google, I sensed, smart is still the starting point. That is, perhaps, its greatest miracle.

I come away from writing this book, *What Would Google Do?*, visiting the company, and meeting and getting to know more of its executives and employees as a fan—a fanboy, we'd say. But I'm not uncritical. In *WWGD?*, I was unhappy with Google's lack of transparency in revealing how much revenue it shares with sites on which its ads appear. It has since revealed that it gives 68 percent to publishers for content ads, and 51 percent for search ads. I also complained about Google's China policy, doing the government's bidding and censoring searches on such topics as Tiananmen Square and Falun Gong. Google has since reversed its policy, after also revealing that it had been attacked by hackers in China who, it's intimated, worked at the behest of the Chinese government. So Google has erased two of my complaints against it.

In the meantime, Google has entered into a deal with Verizon to suggest that the U.S. Federal Communications Commission slice the internet in two, assuring network neutrality in the wired but not the wireless net. But as I just said, soon there will be no clear line between the two networks as our devices—many powered by Google and supplied by Verizon—travel inside and outside, reaching a single internet through wires or radio waves. I find Google's acquiescence to telecom lobbying to be cynical and disappointing. So I'm not a pure fanboy.

Nevertheless, I still find Google ceaselessly fascinating to watch and analyze, as I try to understand its strategies and its impact on how we do business and even how we think. I do not believe, as author Nick Carr has fretted, that Google is making us stupid, nor do I think the internet is changing our brains. But it is certainly changing our habits, letting us assume now that any bit of information is only a search away—and if it's not, there's something wrong with the company or government that's holding it from us. Google is, individually and collectively, our memory. It gets us where we're going. It can understand us in almost any language and translate what we say into almost any other. In perhaps its greatest gift to our productivity, it not only delivers our email but prioritizes it for us.

My study of Google continues. The best way to keep comparing notes together about the company and its era is to join me on Leo Laporte's podcast, *This Week in Google* (twit.tv/twig), with our cohost Gina Trapani, where every week we discuss and analyze life and work in the cloud.

I hope you'll also continue to let me know how you've applied the lessons of Google to your endeavors—where it works and where it doesn't. You can still do so at my blog, Buzzmachine.com, and now via Twitter (@jeffjarvis). And I'd be honored if you'd take a look at my next book, *Public Parts: How Sharing in the Digital Age Improves the Way We Work and Live.* It is about the benefits of openess and the limits of privacy. But like *WWGD?*, it is also about the inevitability of change and the challenges and opportunities it brings. We do live in interesting times.

Jeff Jarvis
New York

Acknowledgments and disclosures

First and foremost, I must thank my blog friends—those who have read, commented on, and linked to Buzzmachine.com—for their invaluable, insightful, and generous help with this book. They inspire and teach me. They correct and challenge me. They give me ideas and push mine. Those friends are too numerous to name. I am grateful to them all.

I am grateful to my editor, Ben Loehnen, for every time I cursed him (as in, "Damn, he's right"). Even as I questioned the old means of publishing, he proved its value with his intelligent, perceptive, and always-encouraging editing. And my publisher, Collins, surprised me with its openness to finding new ways online. (When it came to the digital strategy for this book, they said I was the one who wasn't being brave enough.) At the Collins Publishing Group, I thank Carla Clifford, Hollis Heimbouch, Larry Hughes, Matt Inman, Angie Lee, Shawn Nicholls, Carolyn Pittis, Catherine Barbosa-Ross, Steve Ross, and Margot Schupf for their work to make this book a success.

I also thank my agent, Kate Lee of ICM—the first agent in the industry to respect blogs as sources of talent and ideas. Kate patiently tolerated my ideas and pushed for better ones until we clicked on *What Would Google Do?*

My family could not have been more wonderful through the process of writing this book. My brilliant and beautiful wife, Tammy, tolerated my hours and travels and nerves and made it possible for me to write. My son, Jake, showed me the way to the future. My daughter, Julia, gave me an example as a writer in the family. They tolerated me, too. My parents,

Joan and Darrell Jarvis, and sister, the Rev. Cynthia Jarvis, encouraged me to be a writer and never pointed out that I was only twenty-four years late in meeting my life goal of publishing a book.

Various employers and colleagues generously enabled me to blog and learn digital ways and I thank them. They include Dean Steve Shepard and Associate Dean Judy Watson of the City University of New York; Steve Newhouse of Advance.net; Jim Willse of the Star-Ledger; Alan Rusbridger, Emily Bell, and the editors of Media Guardian at the Guardian; and Upendra Shardanand and Tom Tercek of Daylife. I also thank the editors of BusinessWeek for assigning me reporting that contributed to this book.

I want to thank Peter Hauck, Margaret Kimble, Scott Karp, Clay Shirky, David Weinberger, Doc Searls, Jay Rosen, Rishad Tobaccowala, Fred Wilson, Paulo Coelho, Paula Bracconot, Gary Vaynerchuk, Edward Roussel, Tom Evslin, Seth Godin, Craig Newmark, Samir Arora, Marc Benioff, Chris Bruzzo, Peter Osnos, Jim Louderback, Mark Zuckerberg, Dave Winer, Umair Haque, Martin Nisenholtz, Jeffrey Rayport, Andrew Heyward, Kevin Rose, David Cohn, Dave Morgan, Nick Denton, Scott Heiferman, Chris Anderson, Steven Johnson, Ken Layne, Matt Welch, Caterina Fake, Stewart Butterfield, Bob Garfield, Jimmy Wales, Joan Feeney, Bob Wyman, Will Richardson, Andrew Tyndall, Rick Segal, Bonnie Arnold, Tim O'Reilly, Henry Copeland, Marcel Reichert, Stephanie Czerny, Jochen Wegner, Hubert Burda, Wolfgang Blau, Claudia Gonzalez Gisiger, the World Economic Forum, the Aspen Institute, Lionel Menchaca, Richard@Dell, Michael Dell, and Dell itself.

Note that I am not thanking Google. I am grateful for Google's existence, its lessons, and its inspiration—not to mention Marissa Mayer's quotable advice online. But I want to note that I did not seek access to Google for this book because I wanted to judge it and learn from it at a distance. My admiration of Google, then, does not spring from any relationship with the company but from its incredible example.

Now, in the spirit of transparency and full disclosure:

I have worked for and with many of the organizations I've written about here, including the City University of New York Graduate School of Journalism, the Guardian, Daylife, the New York Times Company, About.com, Advance Publications, Time Warner, Denuo, News Corp., and Burda.

I own stock in various of the companies I've written about, including Google (which I bought as I was finishing research so I would follow its fortunes from a different viewpoint; as I write this, in the depth of the financial crisis, my investment is off about 30 percent), Time Warner, Apple, Amazon, Sirius XM, and Microsoft. I have small investments in startups including Covestor and 33Across and have served on the board of Publish2. At various times, I have advised startups including Technorati, Outside.in, and Meetup.

I receive revenue on my blog from various advertisers via Google AdSense and BlogAds.

I will keep an updated list of disclosures on my blog at buzzmachine .com/about-me.

And finally, thank you for reading my book.

Index

About.com, 41, 88–89
Abrahamson, Shaun, 206
abundance, 3, 130–31
AdSense, 5–6, 33, 58–59
advertising, 5, 29–30, 149–50
    abundance of, 58–59
    in auction marketplace, 69
    book publishing and, 140–41
    data and, 146
    eliminating, 46
    links with, 28
    newspaper, 125–26
    targeted, 179–80
    telecommunications and,
        171
    ubiquity of, 145
advertising agencies
    abundance and, 58
    costs of, 147
    customers as, 46–47
    as networks, 151
    ubiquity of, 145
affiliate programs, 38, 160–61

airlines, 81, 182–86, 244
Albrecht, Alex, 132
Amazon, 4, 31, 160–61, 246
    communities and, 49
    print industry and, 71–72
ambient intimacy, 234
American Girl, 52
analytical reasoning, 217
Anderson, Chris, 63, 79, 233
antitrust, 75, 100
    MLS and, 186
AOL, 5, 50, 80–81, 249
APIs. *See* application programming
    interfaces
Apple, 14–15, 93–94, 226–28, 249
    advertising and, 150
    honesty and, 96
application programming interfaces
    (APIs), 127
Aptera Motors, 175, 193
Arnold, Bonnie, 84
Arrington, Michael, 107
attention, 240

AT&T Worldnet, 30
auction marketplace, 69
automobile industry, 110, 172–77, 246

The Baby Name Wizard (Wattenberg), 233
Baker, Stephen, 159
Ball, James, 205–6
banking, 195–98
Barlow, John Perry, 239
Barton, Rich, 80
Bebo, 49–50
Benioff, Marc, 62
Best Buy, 244
beta versions, 93–95
Betaworks, 193
Bezos, Jeff, 4, 71–72
Biggs, Peter, 146
BlogAds, 50
Blogger, 24–25, 43
Bloglines, 15
Blogpulse, 20
blogs, 22–23
    advertising and, 149–50
    collaboration and, 25, 99
    Comcast and, 168
    Dell and, 12–19
    interacting with, 23
    lawyers and, 222–23
    listening to, 15–16
    9/11 and, 24–25
    profit from, 55–56
    restaurants and, 155
    search engines, 20
    VC and, 190
    video, 157
Blurb.com, 73
BMW, 174
Bono, 163

book publishing, 73–74, 104, 136–43
Bowling Alone (Putnam), 50
Bracconot, Paula, 142
branding, 45, 149
Brilliant, Larry, 163
Brin, Sergey, 85
    on energy, 163–65
    on evil, 99
    Gore *v.*, 217
Bruzzo, Chris, 61–62
Bubblegeneration.com, 64
Buckmaster, Jim, 116–17
Burda, 29–30, 128, 180
Burnham, Brad, 47
Bush, George W., 92
BusinessWeek, 14
Butterfield, Stewart, 45, 89
Buzz, 247, 248
Buzzmachine, 55–56

CafePress.com, 180
Calacanis, Jason, 60
Caravan Project, 140
Carlin, George, 70
Carr, Nicholas, 235, 250
cash flow, innovation *v.*, 110
cashless society, 198
censorship, 99–100, 219, 237, 250
centralization, 27–28
Cerf, Vint, 232
certification, 214
Chicago Tribune, 124
China, 99–100, 105–6, 250
Chowhound.com, 155
Chrome browser, 140, 169
CleverCommute, 188
Clickable, 190

clicks, 28, 66
The Cluetrain Manifesto (Locke, Searls & Weinberger), 3, 82, 96–97, 149
c, mm, n hydrogen car, 175
CNN, 105, 134–35
Coase, Ronald, 151
Coelho, Paulo, 121, 141–43, 240
colas, 178
The Colbert Report, 96
Colbert, Stephen, 96, 136
Cole, Jeffrey, 125
collaboration, 98–99
  Apple and, 226
  blogging and, 25
  with customers, 3–4
  entertainment and, 135
  with government, 219
  links and, 27
  newspapers and, 127–28
  ownership *v.*, 28
Comcast, 107, 167–68
Comedy Central, 136
commodification, 67–68
communication, 217
  customers with, 16–17
  direct, 25
communities
  on airlines, 183
  automobile industry and, 173–74
  elegant organization and, 48–53
  health care and, 200
  restaurants and, 155–56
complaints, 236–37
construction, 225
consumer products, 177–81

consumers
  focus on, 146
  input of, 87–88
content
  commodification of, 67–68
  free, 76–80
  Glam and, 29–30
control
  Apple and, 226
  customer, 3, 11–12
  trust *v.*, 82–83
conversations, 96–97
Cork'd, 159
corporate value, 27–28
corrections, 91
Covestor, 197
craigslist, 31, 38–39, 116–18
  newspaper ads *v.*, 148
Cramer, Jim, 36, 79, 157–58
Cranley, Chris, 205
Crawford, Colin, 70–71
creation, 239–40
credibility, 91
credit crisis, 197
CRM. *See* customer relationship management
crowdsourcing, 156
Culture and Society (Williams, R.), 63
customer relationship management (CRM), 201
customers
  as ad agency, 46–47
  collaborating with, 3–4
  communication with, 16–17
  control of, 3, 11–12
  input, 89–90
  partnering with, 22–23
  retailers and, 161

customers (*continued*)
   trust in, 83–84
   worst, 20–22
customer satisfaction, 167
customer service
   Dell and, 12–15, 18–19
   Google and, 170
   as marketing, 47
   Newmark and, 117
customization, 179

The Daily Show, 95–96
Dale, Iain, 132
data, 88
   advertising and, 146
   health care and, 199–200
   restaurants and, 153–54
   retailers and, 159
Davenport, Thomas H., 111
Daylife, 35, 39, 50, 126
Day, Peter, 113–14
Dean, Howard, 51
Declaration of Independence of
   Cyberspace, 239
Dell, 12–19, 46, 50
Dell'Abate, Gary, 131
Dell, Michael, 14–15, 17–18, 98,
   113
democracy, 238
Denton, Nick, 55, 92
Denuo, 145–46
design
   automobiles and, 172–73
   collaboration and, 98–99
   open, 173
   simplicity and, 116
dhanaX.com, 196
Digg, 4, 86, 126, 128, 132, 247
Diggnation, 132–34

digital equity, 71
digital publishing, 138–39
DIO. *See* Do It Ourselves
Direct2Dell, 17
directory assistance, 78
distributed thinking, 36–39,
   126–27
distribution, 123, 134
doctors, 203
Doerr, John, 165
Do It Ourselves (DIO), 207
domain registration, 170
DonorsChoose.org, 196–97
DoubleClick, 5
Dubner, Stephen J., 75
Dunbar number, 232

eBay, 160
   JetBlue and, 184
   Skype and, 31
   Wal-Mart *v.*, 54–55
Edelman, Richard, 223
education, 104, 210–17
Edwards, D'Wayne, 112
Elberse, Anita, 63
Electronic Frontier Foundation, 239
energy, 162–65
entertainment, 130–36
Entertainment Weekly, 112–13
environment, 126, 162–63
Epicurious.com, 154, 179
Epstein, Daniel A., 148
e-readers, 139–40
Estrada, Joseph, 106
ethics
   of privacy, 232–33
   of publicness, 45
   of transparency, 97
Etsy, 31, 160

European Union, 35
Everyblock, 34
Everything's Miscellaneous
    (Weinberger), 82, 137
evil, 99–102
Evslin, Tom, 30–31
exhibitionism, 234
The Experimental Witch (Coelho),
    142–43
experimentation, 217
eyeballs, clicks *v.*, 66

Facebook, 4, 20–21, 48, 126, 248
    automobile industry and,
        173–74
    Causes application, 196–97
    Google and, 101, 249
    government and, 220–21
    mistakes and, 94–95
    as platform, 34–35
    politics and, 51
    trust and, 85–86
Fake, Caterina, 45, 89
FARC (Revolutionary Armed Forces
    of Colombia), 220
fashion, 103–4, 180
Fast Company, 15
Federal Communications
    Commission (FCC), 131, 166,
    219, 250
Federated, 55
financial meltdown of 2008, 69, 245
First Amendment, 237
Flickr, 21, 45
    communities and, 50
    customer input and, 89–90
Flixwagon.com, 105
FoodBlogBlog, 155
forgiveness, 232

fragmentation, 63, 65
fraud, 205–6
Freakonomics (Levitt & Dubner), 75
free
    book publishing and, 141
    as business model, 76–80
Freedom of Information Act, 218
Fresh Direct, 179
Friedman, Jane, 141–42
Friedman, Thomas, 165
friendships, 231–32

G4, 132
Galant, Debra, 127
Garfield, Bob, 150–51, 167–68
Gawker Media, 55, 92
Generation Google, 7, 231
GetSatisfaction.com, 47
gift economy, 59–63
Glam, 29–30
Gmail, 6, 78–79, 168–69, 247
Godin, Seth, 57, 204–5
Gonzo Marketing (Locke), 149–50
Google
    ambitions of, 121
    antitrust inquiry of, 100
    customer service and, 170
    economy, 68–69
    embedding, 6
    Facebook and, 101, 249
    fooling, 43
    growth of, 69
    home page, 115
    Justice Department investigation
        of, 6
    links and, 27
    media revenue plan, 143–44
    platforms of, 33
    success of, 5–6

Google Analytics, 33
Google Apps, 168
Google Calendar, 33, 168
Google Checkout, 198
Google Docs, 33, 168
Google Groups, 33
Google Health, 200–201
Googlejuice, 42–45, 245
    New York Times and, 78
    Vaynerchuk and, 158
Google Maps, 33–34, 168
    embedding, 6
    real estate and, 188
Google News, 39, 94, 126
Google.org, 162–65
Gore, Al, 163, 217
government, 217–21
Gross, Bill, 175, 193–94
growth, 31–32

Haass, Richard, 237
hacking, 201–2
Haque, Umair, 64, 74, 101–2,
    237
Hatt, Bertil, 204
health care, 199–203, 208
Heiferman, Scott, 206–7
Here Comes Everybody (Shirky), 50,
    60, 151, 237
Heyward, Andrew, 37
Holovaty, Adrian, 34
Holtzbrinck, 193
home pages, 115
honesty, 95–97
Hot, Flat, and Crowded
    (Friedman, T.), 165
Hourihan, Meg, 25
Huack, Peter, 37
Huffington, Ariana, 124

Hughes, Chris, 51
Hulu, 135
Hunter, Dick, 18–19

Icerocket, 15, 20
ICQ, 31–32
Idealab, 175, 193
Ideas platform, 62
IdeaStorm, 17
identity, 233–34
    business, 80–81
Ikea, 140
incubators, 193
Indeed.com, 39
inefficiency, 74, 128–29
InnoCentive, 113–14
innovation, 111–14
    cash flow v., 110
    newspapers and, 129–30
Institute for the Future of the Book, 138
insurance, 203–9
interestingness, 89–90
iPhone, 51, 249
"Is Google Making Us Stupid?"
    (Carr), 235, 250
iterations, 93–94
iTunes, 135
iVillage, 29
Iyer, Bala, 111

Jarvis, Mark, 47
JetBlue, 184
Jobs, Steve, 226–28
    honesty and, 96
Johnson, Charles, 92
Johnson, Steven, 191
journalism
    online, 26
    stagnancy of, 110, 245, 249

transparency and, 92
trust and, 86
Justice Department, 6, 186

Kalsey, Adam, 19
Kaplan University, 217
Karp, David, 192
Kelly, Kevin, 138
Kennedy, Jim, 193
Kindle, 72, 139–40
Kitchen Nightmares, 154
Kiva.org, 196
Knol, 80
Kodak, 81
Kurnit, Scott, 41

Lafley, A.G., 89, 91, 112
lawyers, 222–25
Layne, Ken, 92
leadership, 217
Lehrer, Brian, 128
Lehrer, Tom, 216
Lent, Gregory, 207
Levitt, Steven D., 75
Lexis, 224
link economy, 124–25
LinkedIn, 49
links
   with ads, 28
   SEO and, 44–45
Linux, 17
listening, 15–16, 87–90, 128
LittleGreenFootballs, 92
live, 105–6
LiveMocha, 215
LiveWorld, 87
Loanio, 196
Locke, Christopher, 3, 82, 96–97,
   149–50

LonelyGirl15, 135
The Long Tail (Anderson), 63, 79
Los Angeles Times, 55, 86–87
Louderback, Jim, 129, 133–35
Lulu.com, 73
Lutz, Bob, 173

Madonna, 93–94
magazines, 70–71, 77
Mahalo, 60
MAIF. See Mutuelle Assurance
   Instituteur France
Malseed, Mark, 114–15
map mashups, 33
marketing
   customer service as, 47
   mass, 149
   piracy as, 141–42
mass market
   decline of, 3
   niches v., 63–67
Masters, George, 150
Mayer, Marissa, 31, 34, 87–88
   innovation and, 111–12
   on iterations, 93–94
McCain, John, 136
McIntosh, Neil, 214
Mechanical Turk, 72
media
   buying, 150
   fragmentation of, 65
   mistrust of, 83
   revenue plan, 143–44
Meetup, 50
Menchaca, Lionel, 17–18
Meredith, Scott, 139
meritocracies, 99
Meyer, Philip, 125
microloans, 196

Microsoft, 16, 35
middlemen, 73–76
military, 225
Miller, Robert, 138–40
The Mining Company, 41
mistakes, 91–93
MLS. *See* multiple-listing service
M&Ms, 179
mobs, 106–8
Moore, Gordon, 133
Moore's Law, 133
Motley Fool, 197
movies, 133–35
multiple-listing service (MLS), 75, 186
Murdoch, Rupert, 129
music industry, 110
Mutuelle Assurance Instituteur
    France (MAIF), 204, 208
MyStarbucksIdea.com, 60–62

National Association of Broadcasters,
    167
"The Nature of the Firm" (Coase), 151
Netflix, 160
Netscape, 67
networks, 4
    advertising agencies as, 151
    on airlines, 183–85
    insurance as, 206
    joining, 27–32
    open, 29–30
    on platforms, 32–33
    social, 231
    specialization and, 154
    universities and, 212–13
network theory, 28
"New Economics of Media"
    (Haque), 64
Newmark, Craig, 38–39, 47, 116–18

newness, 146
newspaper ads, 38–39, 125–26
    craigslist *v.*, 148
    decline of, 76
newspapers, 3, 123–30, 245, 250
New York Angels, 193
New York Times, 41
    API of, 127
    free content and, 78
niches
    mass market *v.*, 63–67
    newspapers and, 129
Nike, 112
9/11, 24–25
Nisenholtz, Martin, 37
Nocera, Joe, 100
*Nothing Sacred: The Truth About
    Judaism* (Rushkoff), 226
NPR, 127
The Numerati (Baker), 159

Obama, Barack, 51, 220–21, 232
*101 Wines Guaranteed to Inspire,
    Delight and Bring Thunder to
    Your World* (Vaynerchuk), 157
OnStar, 177
Oodle.com, 39
openness, 236
    Apple and, 227
    ownership *v.*, 4
    PR and, 224
open-source, 59–63
    Apple and, 227
    automobile industry and, 174–75
    Coelho and, 142–43
    fashion, 180
    religion and, 226
    restaurants and, 154
Open-Source Judaism, 226

O'Reilly, Tim, 79
organization, elegant, 48–53
  airlines and, 182
  newspapers and, 129
Orkut, 50
Osnos, Peter, 140
Outside.in, 191
ownership, 4, 28
Owyang, Jeremiah, 172–73

Page, Larry, 85, 248
  on environment, 162–64
  on evil, 99
  Gore *v.*, 217
  on mistakes, 94
  telecommunications and, 166
PageRank, 85
partnering, 22–23, 151
passion, 217
passivity, 116–18
patents, 224
Paterson, David, 96
PatientsLikeMe, 200
PayPal, 86, 197–98
Peapod, 179
peer-to-peer lending, 176–77
personal liberty, 238
personal political pages (PPPs),
  219
Picasa, 33, 81, 193
piracy, as marketing, 141–42
platform(s), 32–36
  craigslist as, 117–18
  newspapers and, 126–27
Platial.com, 33–34
politics, 51, 217–21
  trust and, 83
Pope, Ivan, 206
populism, 84–85

pornography, 225
PornTube, 225
portfolios, 214
post-scarcity economy, 57–59
Potts, Mark, 56
PowerPoint, 64
Poynter, Don, 137
PPPs. *See* personal political pages
Prezvid.com, 37
print-on-demand, 140
Prius, 164, 175
privacy, 232–33
Procter & Gamble, 89, 91
Prosper.com, 86, 195
PR. *See* public relations
Publicis Groupe Media, 145–46
publicness, 45, 232–33
public relations (PR), 222–25
Putnam, Robert, 50

Qik.com, 105
quality, 27, 147, 228

Rademacher, Paul, 33
Ramsey, Gordon, 154
RateMyTeachers.com, 215
Rather, Dan, 92
Rayport, Jeffrey, 37, 211–12
RE<C, 162
real estate, 75, 186–88
RechargeIT, 164–65
Reichelt, Leisa, 234
religion, 226–28
restaurants, 153–57
retailers, 157–61
revenue
  digital publishing and, 138–39
  growth *v.*, 31
  for media, 143–44

revenue (*continued*)
  movie, 134
  side door, 146
Revision3, 130, 132–34
Revolutionary Armed Forces of
    Colombia. *See* FARC
The Revolution Will Not Be Televised
    (Trippi), 238
Rheingold, Howard, 106
Richardson, Will, 211
Rose, Kevin, 4, 132, 134
Rosenberg, Jonathan, 217
Rosen, Jay, 134–35
Roussel, Edward, 123
Rubel, Steve, 223
Rusbridger, Alan, 126
Rushkoff, Douglas, 226
Ryanair, 79
Ryan, Pat, 64

Salesforce.com, 62
Sandberg, Sheryl, 94
SANS, 180
scale, 54–57
Schmidt, Eric, 248
  on Carr, 235
  on Gmail, 6
  on home-page sponsors, 36
  on mistakes, 94
  on mobile market, 79
Scion, 174
Scoble, Robert, 150
search-engine optimization (SEO),
    41–42, 44–45
search engines, 5, 20
Searls, Doc, 3, 82, 96–97, 149,
    170
  VRM and, 201–2
secrets, 97
Seed Camp, 193

Seesmic.com, 142
Segal, Rick, 15, 95
self-publishing, 73
self-searches, 20
Semel, Terry, 81
SEO. *See* search-engine optimization
Sequoia Capital, 189
Shardanand, Upendra, 35
Shirky, Clay, 50, 60, 151, 191–92,
    197, 235–36, 237
Silverman, Dwight, 13
simplicity, 114–16, 236
SimplyHired.com, 39
Sirius Satellite Radio, 131
Skype, 31, 50
Smart Mobs (Rheingold), 106
Smith, Quincy, 38
Smolan, Rick, 140
social business, 158
social graph, 49
socialization, 211–12
social-media, 172–73
social responsibility, 47
social web, 51
Sorrell, Martin, 42
Sourcetool.com, 100
specialization, 26–27, 154
speed, 103–4, 105–6
Spitzer, Eliot, 96
splogs, 43
Starbucks, 60–62
Stern, Howard, 95, 131–32
Stewart, Jon, 95–96
StudieVZ, 50
Supreme Court, 225
Surowiecki, James, 88

talent, 146, 240
Tapscott, Don, 113, 151, 225
targeting, 151, 179–80

teaching, 193, 214–15
teamwork, 217
TechCrunch, 107, 192
Technorati, 15, 20
TechTV, 132
telecommunications, 165–71
Telegraph Media Group, 123
television, 84
    cable, 167
    decline of, 65–66
    listings, 109–10
    networks, 135
Television Without Pity, 135
Tesco, 179
Tesla Motors, 175
testing, 214
Threadbanger, 180
Threadless, 57
TimesSelect, 78
Time Warner, 80–81
Tobaccowala, Rishad, 114, 121–22,
    145–48, 151, 177
    on Apple, 228
toilet paper, 180–81
TomEvslin.com, 31
Toto, 181
Toyota, 174–75
transparency, 83, 97–98
    journalism and, 92
    PR and, 223
Tribune Company, 129
Trippi, Joe, 238
trust, 74, 170
    control v., 82–83
    in customers, 83–84
Tumblr, 192
Turner, Ted, 134
TV Guide, 109–10
20 percent rule, 111, 114
23andMe, 205

Twitter, 20, 126, 247
    Dell and, 46
    mobs and, 107
    real time and, 105–6
Tyndall, Andrew, 220

Union Square Ventures, 30
University of Phoenix, 217
Updike, John, 138

The Vanishing Newspaper (Meyer),
    125
Vardi, Yossi, 31–32
Vaynerchuk, Gary, 107, 157–61
VC. See venture capital
vendor relationship management
    (VRM), 201–2
venture capital (VC), 189–95
Vershbow, Ben, 138
Virginia Tech University, 105
Virgin Money, 197
Virtual Law Partners, 223
Vise, David A., 114–15
VRM. See vendor relationship
    management

Waghorn, Rick, 56
Wales, Jimmy, 60, 87
Wall Street Journal, 129
Wal-Mart, 54–55, 101
Washlet, 181
Wattenberg, Laura, 233
Weinberger, David, 3, 82, 96–97,
    137, 149, 232
Westlaw, 224
widgets, 36–37
Wikia, 60
Wikileaks.org, 92–93
Wikinomics (Tapscott), 113, 151,
    225

Wikipedia
    communities and, 50
    growth of, 66
    mistakes in, 92–93
    open-source and, 60
    speed of, 106
wikitorials, 86–87
Williams, Evan, 105–6
Williams, Raymond, 63
Wilson, Fred, 35, 176, 189–92, 225, 237
Wine.com, 158
WineLibrary.TV, 157
The Winner Stands Alone (Coelho), 142
Wired, 33
wireless access, 166
    airlines and, 182–83
wireless spectrum, 166
The Wisdom of Crowds (Surowiecki), 88
The Witch of Portobello (Coelho), 142–43
WNYC, 128

Wojcicki, Anne, 205
Wolf, Maryanne, 235
World Economic Forum, 48, 113
Wyman, Bob, 211

Yahoo, 5, 36, 58
    China and, 99–100
    communities and, 50
Yang, Jerry, 36
Y Combinator, 193
youth, 191–94, 212
YouTube, 6, 20, 33, 37

Zappos, 161
Zara, 103–4
Zazzle, 180
Zell, Sam, 129
zero-based budgeting, 79–80
Zillow, 75, 80, 187
Zipcar, 176
Zopa, 196
Zuckerberg, Mark, 4, 48–53, 94–95